THE MOLLIFICATION METHOD
AND THE NUMERICAL SOLUTION
OF ILL-POSED PROBLEMS

THE MOLLIFICATION METHOD AND THE NUMERICAL SOLUTION OF ILL-POSED PROBLEMS

DIEGO A. MURIO
Department of Mathematical Sciences
University of Cincinnati

A Wiley-Interscience Publication
JOHN WILEY & SONS, INC.
New York · Chichester · Brisbane · Toronto · Singapore

Library of Congress Cataloging in Publication Data:

Murio, Diego A., 1944–
    The mollification method and the numerical solution of ill-posed
problems / by Diego A. Murio.
       p. cm.

    "A Wiley interscience publication."
    Includes bibliographical references and index.
    ISBN 0-471-59408-3
    1. Inverse problems (Differential equations)—Improperly posed
problems. 2. Inverse problems (Differential equations)—Numerical
solutions. I. Title.
QA377.M947 1993
515'.353—dc20   93-163

To the musicians in my life:
Diego S., Veronica, and Francis.

CONTENTS

PREFACE

During the last 20 years, the subject of ill-posed problems has expanded from a collection of individual techniques to a highly developed and rich branch of applied mathematics. This textbook essentially builds, from basic mathematical concepts, the understanding of the most important aspects of the numerical treatment of the applied inverse theory.

The subject has grown—and continues to grow—at such a fast pace that it is impossible to offer a complete treatment in an introductory textbook and all that can be done is to discuss a few important and interesting topics. Inevitably, in making the selection, I have been influenced by my own interests which, on the other hand, allowed me the pleasure to write about the particular problems with which I am most familiar. This book is intended to be a self-contained presentation of practical computational methods which have been extensively and successfully applied to a wide range of ill-posed problems. The nature of the subject demands the application of special mathematical techniques—rarely seen in typical science courses and strange to normal engineering curricula—with which it is initially difficult to relate the steps of a calculation with the more classical concepts of stability and accuracy. This book is intended to solve the problem by giving an account of the theory that builds from the phenomena to be explained, keeping everything in as elementary a level as possible, making it useful to a wide circle of readers.

The primary goal of this book is to provide an introduction to a number of essential ideas and techniques for the study of inverse problems that are ill posed. There is a clear emphasis on the mollification method and its multiple applications when implemented as a space marching algorithm. As such, this book is intended to be an outline of the numerical results obtained with the mollification method and a manual of various other methods which are also used in arriving at some of these results. Although the presentation concentrates mostly on problems with origins in mechanical engineering, many of the ideas and methods can be easily applied to a broad class of situations.

This book—an outgrowth of classes that the author has taught at the University of Cincinnati for several years in the Seminar of Applied Mathematics—is organized around a series of specific topics aimed at upper-level undergraduates and first-year graduate students in applied mathematics, the sciences, and engineering. It may be used as a primary test for a course on computational methods for inverse ill-posed problems or as a reference work for professionals interested in modeling inverse phenomena in general.

The treatment is strongly computational, with many examples and exercises, and truly interdisciplinary. There is more than enough material in the book to be covered in a semester or two-quarters-long course. Although all the problems considered are physically motivated, a knowledge of the physics involved is not essential—but always very useful!—for the understanding of the mathematical aspects of these problems. It has been my experience, after teaching this material several times, that the subject is most appreciated by the students when they write programs of their own and see them work. This is the main reason for having computational exercises in the book. An "experimental" approach to numerical modeling—defining and carefully testing the new numerical methods on simple problems with known solutions, before attempting to formally prove their stability and accuracy—is often the best way to proceed. The students are advised to work on as many exercises as possible in each chapter, as this is the best way to learn the material and to check if students master the subject.

Chapter 1 begins with the classical topic of numerical differentiation as an inverse problem. The mollification method is then introduced and a thorough discussion of its numerical implementation follows.

In Chapter 2 we investigate Abel's integral equation—another classical problem of mathematical physics—and four different methods are developed and placed in their proper computational framework.

Chapter 3—the main thrust of the book—is devoted to the one-dimensional inverse heat conduction problem (IHCP). The discussion of several mathematical models and their pertinent numerical algorithms for the approximate determination of the unknown transient temperature and heat flux functions is developed here.

Chapter 4 on the two-dimensional IHCP is presented as an integral part of the text. It is in this topic area where the reader can find the most prolific and challenging source of new problems.

Chapter 5 contains three independent applications for space marching solutions of the IHCP: identification of boundary source functions and radiation laws, numerical solutions of the Stefan problem and inverse Stefan problem, and determination of the initial temperature distribution in a one-dimensional conductor from transient measurements at interior locations.

Chapter 6 illustrates further applications of stable numerical differentiation techniques ranging from the determination of forcing terms in systems of ordinary differential equations to the identification of transmissivity coefficients in linear and nonlinear elliptic and parabolic equations in one space dimension.

Appendix A offers a selected overview of the essential mathematical tools used in these lectures. It should constitute a genuine aid for nonmathematicians.

Finally, Appendix B contains an up-to-date citing of the literature related to the IHCP. It might certainly be of value to graduate students and researchers interested in the subject.

DIEGO A. MURIO

ACKNOWLEDGMENTS

I would like to acknowledge the contribution of several of my graduate students—Constance Roth, Lijia Guo, and Doris Hinestroza—who helped me test and debug portions of the original manuscript under diverse circumstances. I feel particularly indebted to Carlos E. Mejía who read the entire text and made many valuable suggestions while still working on his Ph. D. Dissertation.

Finally, I wish to acknowledge the following organizations for permission to reproduce the indicated tables and figures: Society for Industrial and Applied Mathematics (SIAM), Figures 2.6 and 2.7; IOP Publishing Ltd., Figures 4.1a, 4.1b, 4.2a, and 4.2b and Tables 4.1 and 4.2; Pergamon Press, Figures 5.1 to 5.7, 5.10 to 5.13, 6.3, 6.6, 6.7b, and 6.8 and Tables 5.1, 5.2, and 6.1 to 6.3.

D. A. M.

1

NUMERICAL DIFFERENTIATION

In several practical contexts, it is sometimes necessary to estimate the derivative of a function whose values are given approximately at a finite number of discrete points. It is easy to imagine many different situations—mostly involving integral equations and ordinary and partial differential equations—related with the question of numerical differentiation of measured (noisy) data. Several interesting applications of this basic problem will be investigated in the following chapters.

1.1 DESCRIPTION OF THE PROBLEM

In order to gain some insight on the underlying principles, let us analyze first the ideal situation where we seek an approximation to the derivative function $f'(x)$ under the assumption that the exact (errorless) data function $f(x)$ is sufficiently smooth on a given interval $[a, b]$. For example, if we assume that $f \in C^3([a, b])$—third derivative continuous on $[a, b]$—and satisfies the uniform bound

$$\|f'''\|_{\infty, [a, b]} = \max_{a < x < b} |f'''(x)| \le M_3,$$

it is possible to approximate the derivative $f'(x)$ by the centered difference

$$D_0 f(x) = \frac{f(x + h) - f(x - h)}{2h}, \qquad a \le x - h < x < x + h \le b.$$

We can estimate the truncation error by Taylor's series. In fact, we have

$$f(x + h) = f(x) + hf'(x) + \frac{h^2}{2}f''(x) + \frac{h^3}{6}f'''(s_1), \qquad x < s_1 < x + h,$$

$$f(x - h) = f(x) - hf'(x) + \frac{h^2}{2}f''(x) - \frac{h^3}{6}f'''(s_2), \qquad x - h < s_2 < x.$$

Subtracting the last two expressions and dividing by $2h$, we get

$$D_0(x) = f'(x) + \frac{h^2}{6}(f'''(s_1) + f'''(s_2))$$

and

$$|D_0 f(x) - f'(x)| \le \frac{h^2}{3}M_3 = O(h^2). \tag{1.1}$$

The approximation of the derivative by centered differences—in the absence of noise in the data—is a second-order consistent procedure, that is, as h approaches 0, $D_0 f(x)$ converges to $f'(x)$ with rate proportional to h^2.

However, in real situations we deal with experimentally determined (measured) data that contain errors, systematic and random. The following discussion will assume that any systematic errors have been removed from the supplied data function. Consequently, instead of the ideal data function $f(x)$, we consider a measured data function $f_m(x)$ obtained by adding the random noise function $N(x)$ to $f(x)$. We also assume that the amplitude of the noise is bounded by ε. Thus,

$$f_m(x) = f(x) + N(x), \qquad a \le x \le b,$$

and

$$\|N(x)\|_{\infty, [a, b]} = \max_{a \le x \le b} |N(x)| \le \varepsilon.$$

To illustrate the new situation, consider a smooth but arbitrary ideal data function $f(x)$ and the also very smooth noisy function $N(x) = \alpha \sin(wx)$, $|\alpha| \le \varepsilon$, $w \in \mathbb{R}$, defined on the entire real line. Then

$$f_m'(x) = f'(x) + \alpha w \cos(wx),$$

and for small α and large w, the derivative error

$$f_m'(x) - f'(x) = \alpha w \cos(wx)$$

is greatly amplified for $x \cong \pm k\pi/w$, $k = 0, 1, 2, \ldots$.

In terms of centered differences,

$$D_0 f_m(x) = D_0 f(x) + \frac{N(x+h) - N(x-h)}{2h}$$

$$= D_0 f(x) + \frac{\alpha}{h} \sin(wh) \cos(wx). \tag{1.2}$$

Now suppose that w is chosen, as a function of h, to be $w = \pi/2h$. Hence, for each h,

$$D_0 f_m(x) = D_0 f(x) + \frac{\alpha}{h} \cos\left(\frac{\pi x}{2h}\right)$$

and the second term on the right-hand side—the rounding error—is inversely proportional to the interval of computation h. Therefore, with a decrease in h the rounding error increases. This is indeed the case in general. Even if $N(x)$ represents a random noise variable with amplitude $|N(x)| \le \varepsilon$, we have

$$-\frac{\varepsilon}{h} \le \frac{N(x+h) - N(x-h)}{2h} \le \frac{\varepsilon}{h}$$

and whenever the numerator $N(x+h) - N(x-h)$ is different from 0, the rounding error is greatly amplified for small values of h.

From a more honest computational point of view, we would like to estimate the difference between the "ideal" derivative function $f'(x)$—a complete abstract object numerically—and the actual "computed" value $D_0 f_m(x)$ obtained from the measured data using finite differences. We proceed as follows.

From

$$f'(x) - D_0 f_m(x) = f'(x) - D_0 f(x) + D_0 f(x) - D_0 f_m(x),$$

using the triangle inequality, (1.1) and (1.2),

$$|f'(x) - D_0 f_m(x)| \le |f'(x) - D_0 f(x)| + |D_0 f(x) - D_0 f_m(x)|$$

$$\le \frac{h^2}{3} M_3 + \frac{|\alpha|}{h} |\sin(wh)| |\cos(wx)|.$$

If we consider, as before, the particular values $w = \pi/2h$, the estimate becomes

$$|f'(x) - D_0 f_m(x)| \le \frac{h^2}{3} M_3 + \frac{|\alpha|}{h} \left| \cos\left(\frac{\pi x}{2h}\right) \right|,$$

which strongly suggests choosing h so as to minimize the upper bound for the total error. This seems reasonable since, in general, as h decreases, the truncation error decreases and the rounding error increases.

The upper bound, achievable for some values of x, shows that at least near those points it is not possible to approximate the ideal derivative function $f'(x)$ by centered finite differences as in (1.1); now, as h approaches 0, the error blows up! This result is, of course, totally expected because the known instability of $f'_m(x)$ is inherited by the finite-difference approximation $D_0 f_m(x)$.

The preceding discussion shows that the process of differentiation is such that small errors in the data function might produce large errors in the derivative function, independently of how smooth the error function is. Moreover, the same occurrence is notably true in the context of discrete approximations for numerical differentiation with noisy data. This situation is also encountered in many other problems and justifies the following definition.

DEFINITION 1.1 (Hadamard)   A mathematical problem is said to be well posed if it has a unique solution and the solution depends continuously on the data.

A problem which is not well posed is said to be ill posed. The differentiation problem is an example of the latter.

The conclusion is simple. The application of standard numerical techniques to the process of numerical differentiation might yield nonphysical "solutions." What can be done?

First of all, we need to characterize the source of ill-posedness for the differentiation problem. A new insight can be obtained by performing a Fourier analysis of the process. If $f(x), f'(x) \in L^2(\mathbb{R})$, then $\hat{f}'(w) = iw\hat{f}(w)$. This means that f is not just "a function in $L^2(\mathbb{R})$"; its high-frequency behavior is such that $\|\hat{f}\|$ decreases faster than $\|w^{-1}\|$ as $\|w\| \to \infty$. Now, even if we assume that the noise function $N(x) \in L^2(\mathbb{R})$, there is no reason to believe that the high-frequency components of $\hat{N}(w)$ will be subject to such a rapidly decreasing behavior and we cannot, in general, guarantee that the product $(iw)\hat{N}(w)$ will be in $L^2(\mathbb{R})$.

We conclude that the differentiation problem, in this setting, *is an ill-posed problem in the high-frequency components* and any attempt to stabilize the problem—restore continuity with respect to the data—in order to be successful, must take this consideration into account.

1.2   STABILIZED PROBLEM

Let $C^0(I)$ denote the set of continuous functions over the interval $I = [0, 1]$ with $\|f\|_{\infty, I} = \max_{x \in I}|f(x)| < \infty$.

We consider the problem of estimating in I the derivative $f'(x)$ of a function $f(x)$ defined on I and observed with error. We assume that $f(x)$ is twice continuously differentiable on I. Instead of $f(x)$, we know some data function $f_m(x) \in C^0(I)$ such that $\|f_m - f\|_{\infty, I} \leq \varepsilon$.

Our initial task is to stabilize the differential process. To that end, we first introduce the function

$$\rho_\delta(x) = \frac{1}{\delta\sqrt{\pi}} \exp\left(\frac{-x^2}{\delta^2}\right),$$

the Gaussian kernel of "blurring radius" δ. We notice that ρ_δ is a C^∞ (infinitely differentiable) function that falls to nearly 0 outside a few radii from its center ($\cong 3\delta$), is positive, and has total integral 1.

After extending the ideal data function $f(x)$ and the measured data function $f_m(x)$ to the interval $I_\delta = [-3\delta, 1 + 3\delta]$ in such a way that they decay smoothly to 0 in $[-3\delta, 0] \cup [1, 1 + 3\delta]$ and they are 0 in $\mathbb{R} - I_\delta$, the convolution

$$J_\delta f(x) = (\rho_\delta * f)(x) = \int_{-\infty}^{\infty} \rho_\delta(x - s)f(s)\,ds$$

$$\cong \int_{x-3\delta}^{x+3\delta} \rho_\delta(x - s)f(s)\,ds, \tag{1.3}$$

defines a C^∞ function in the entire real line.

The extensions can be easily accomplished, for instance, by defining

$$f_m(x) = f_m(0)\exp\left\{x^2/\left[(3\delta)^2 - x^2\right]\right\}, \qquad -3\delta \leq x \leq 0,$$

$$f_m(x) = f_m(1)\exp\left\{(x - 1)^2/\left[(x - 1)^2 - (3\delta)^2\right]\right\}, \qquad 1 \leq x \leq 1 + 3\delta. \tag{1.4}$$

$J_\delta f$ is the mollifier of f and δ is the radius of mollification.

Notice that the Fourier transform of $J_\delta N(x)$ is given by

$$(\widehat{J_\delta N})(w) = \hat{\rho}_\delta(w)\hat{N}(w) = \frac{1}{\sqrt{2\pi}}\exp\left(-\frac{w^2\delta^2}{4}\right)\hat{N}(w)$$

and damps those Fourier components of the noise N with wavelength $2\pi/w$ much shorter than $2\pi\delta$; the longer wavelengths are damped hardly at all, as

$$\frac{2\pi}{\omega} < 2\pi\delta \iff 1 < \omega\delta$$

required by our analysis in the previous section. We also observe that if f has compact support K, the mollification function defined by (1.3) does not have compact support and, as such, does not verify Definition A.9. However, for all practical purposes we can consider its "numerical support" to be compact and extended up to approximately 3δ units away from the boundary of K.

Moreover, from the definition, it follows immediately that

$$\frac{d}{dx} J_\delta f(x) = (\rho_\delta * f)'(x) = (\rho_\delta * f')(x) = (\rho'_\delta * f)(x).$$

The following two lemmas are fundamental for our results.

Lemma 1.1 (Consistency)   If $\|f''\|_{\infty, I} \leq M_2$, then

$$\|(\rho_\delta * f)' - f'\|_{\infty, I} \leq 3\delta M_2.$$

Proof. For $x \in I$,

$$(\rho_\delta * f)'(x) = \int_{-\infty}^{\infty} \rho_\delta(x - s) f'(s)\, ds$$

and

$$f'(x) = \int_{-\infty}^{\infty} \rho_\delta(x - s) f'(x)\, ds.$$

Subtracting and using the mean value theorem,

$$|(\rho_\delta * f)'(x) - f'(x)| \leq \int_{-\infty}^{\infty} \rho_\delta(x - s)|f'(s) - f'(x)|\, ds$$

$$\cong \int_{x-3\delta}^{x+3\delta} \rho_\delta(x - s)|f'(s) - f'(x)|\, ds$$

$$\leq 3\delta M_2.$$

Thus,

$$\|(\rho_\delta * f)' - f'\|_{\infty, I} \leq 3\delta M_2. \quad \blacksquare$$

Lemma 1.2 (Stability)   If $f_m(x) \in C^0(I)$ and $\|f_m - f\|_{\infty, I} \leq \varepsilon$, then

$$\|(\rho_\delta * f_m)' - (\rho_\delta * f)'\|_{\infty, I} \leq \frac{2\varepsilon}{\delta\sqrt{\pi}}.$$

Proof. For $x \in I$,

$$\left| (\rho_\delta * f_m)'(x) - (\rho_\delta * f)'(x) \right|$$

$$\leq \int_{-\infty}^{\infty} \left| \frac{d}{dx} \rho_\delta(x - s) \right| \left| f_m(s) - f(s) \right| ds$$

$$\leq \varepsilon \int_{-\infty}^{\infty} \left| \frac{d}{dx} \rho_\delta(x - s) \right| ds = 2\varepsilon \int_0^{\infty} \frac{d}{dx} \rho_\delta(x) \, dx = \frac{2\varepsilon}{\delta \sqrt{\pi}}. \quad \blacksquare$$

Lemma 1.2 shows that attempting to reconstruct the derivative of the mollified data function is a stable problem with respect to perturbations in the data, in the maximum norm, and for δ fixed.

Theorem 1.1 (Error Estimate)   Under the conditions of Lemmas 1.1 and 1.2,

$$\left\| (\rho_\delta * f_m)' - f' \right\|_{\infty, I} \leq 3\delta M_2 + \frac{2\varepsilon}{\delta \sqrt{\pi}}. \tag{1.5}$$

Proof. The estimate follows from Lemmas 1.1 and 1.2 and the triangle inequality. $\blacksquare$

We observe that the error estimate (1.5) is minimized by choosing $\delta =$ $\bar{\delta} = [2\varepsilon / 3M_2 \sqrt{\pi}]^{1/2}$. For this "optimal" choice of the radius of mollification, the error estimate becomes

$$\left\| (\rho_{\bar{\delta}} * f_m)' - f' \right\|_{\infty, I} \leq 2\pi^{-1/4} \sqrt{6M_2} \sqrt{\varepsilon},$$

and we obtain uniform convergence as $\varepsilon \to 0$—as the quality of the data improves—with rate $O(\varepsilon^{1/2})$.

However, in practical computations we have to solve the problem with a fixed upper bound ε, and the choice $\delta = \bar{\delta}$ is impossible because M_2 is not known, in general.

Before discussing the selection of the radius of mollification, let us summarize our approach. This same basic strategy will be used again and again to solve different ill-posed problems and with different methods.

First, we replace the original ill-posed problem of finding f' by the new problem of finding $J_\delta f'$. The new problem is well posed, depends on a parameter $\delta > 0$, and, in the absence of noise in the data, is consistent with the original problem (Lemma 1.1).

Second, in the presence of noise in the data, with a fixed value of the parameter $\delta > 0$, we solve the new problem—not the original one—which is stable with respect to perturbations in the data (Lemma 1.2).

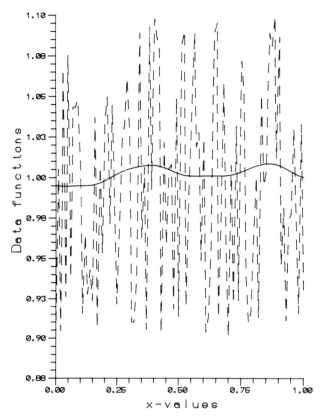

Fig. 1.1   Data function $f_m(x_i) = 1 + \varepsilon \theta_i$ (- - -); mollified data function $J_\delta f_m(x_i)$ (—). $\varepsilon = 0.1$, $\delta = 0.08$, $h = 0.01$.

It is customary to say that the family $J_\delta f'$, $\delta > 0$—satisfying Lemmas 1.1 and 1.2—is a *regularizing family* for the differentiation problem.

Figures 1.1 and 1.2 illustrate a random noise function of amplitude 0.1 added to $f(x) \equiv 1$ in real x space, its Fourier transform, the mollified (filtered) version in the frequency space, and the mollified noise function back in real space, respectively. Observe carefully the relationships among the high-frequency components before and after the filtering procedure.

Remarks

1. It is a simple task to estimate the error between the exact derivative f' restricted to the grid points and the centered-difference approximation $D_0(\rho_\delta * f_m)$. Given that $(\rho_\delta * f_m) \in C^\infty(I)$, if $M_{3,\delta}$ denotes a uniform upper bound for its third derivative, we simply combine estimates (1.1)

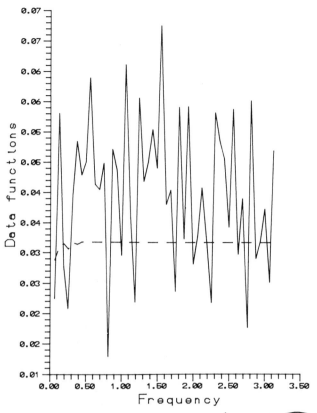

Fig. 1.2   Normalized discrete Fourier transforms: $\hat{f}_m(w)$ (—); $\widehat{(J_\delta f_m)}(w)$ (- - -).

and (1.2) using the triangle inequality to obtain

$$\| D_0(\rho_\delta * f_m) - f' \| \leq \frac{h^2}{3}M_{3,\delta} + 3\delta M_2 + \frac{2\varepsilon}{\delta\sqrt{\pi}}.$$

2. For the analysis of a totally discretized approach of the mollification method, the reader is referred to the exercises in Sec. 1.7.

1.3   DIFFERENTIATION AS AN INVERSE PROBLEM

Consider a Fredholm linear integral equation of the first kind

$$f(x) = \int_0^1 k(x,s)g(s)\,ds, \qquad (1.6)$$

where the kernel function k is square integrable in $[0, 1] \times [0, 1]$ and the only solution of the homogeneous problem

$$0 = \int_0^1 k(x, s) g(s) \, ds$$

is $g(x) \equiv 0$, $0 \le x \le 1$.

Ordinarily, k and g are known functions and we are asked to determine the solution function f so as to satisfy (1.6). So posed, this is a direct problem.

There is, however, an interesting inverse problem that can be formulated. The objective of this new problem is to determine part of the "structure" of the system represented by the prototype Fredholm equation. In our case, we would like to find the function $g(x)$ from experimental information given by the approximate knowledge of the solution of the direct problem $f(x)$.

From the physical point of view, the distinction between a direct (going from g to f) and an inverse problem (going from f to g) is of a phenomenological nature. Actual experiments are associated only with direct problems (cause–effect relationship); there are no physical experiments directly linked to inverse problems. It is this unavoidable and natural physical fact that it is sometimes hidden in the mathematical model. This means that in order to experimentally validate the solution of an inverse problem, it is necessary to use this solution a posteriori, as data for the experiment associated with the direct problem. As an example, consider the solution of the heat equation forward in time with initial (time 0) and boundary conditions as the usual direct problem. There are several inverse problems associated with this direct problem—some of them to be discussed later—but, for the moment, we can restrict our attention to the inverse problem (backward in time) of attempting to recover the initial condition from knowledge of the boundary conditions and the temperature distribution at some time $t > 0$. It is clear that the inverse problem cannot be directly validated by an experiment because it is impossible to physically reverse the flow of time.

The preceding discussion, obviously, does not apply to mathematical modeling. In the much simpler situation of numerical differentiation, we point out that if in (1.6) the kernel k is given by

$$k(x, s) = \begin{cases} 0, & x < s, \\ 1, & x \ge s, \end{cases}$$

then solving for g with f as data is equivalent to differentiating g, that is, $g(x) = f'(x)$.

1.4   PARAMETER SELECTION

In this section we indicate a procedure to uniquely determine the radius of mollification δ as a function of the amount of noise in the data ε, based only on properties of the filtered data function $\rho_\delta * f_m$. In order to proceed formally, we extend $f_m(x)$ to the whole x axis as indicated in Sec. 1.3 and freely use the theory of Fourier transforms in L^2, in particular Parseval's relation (Theorem A.10) and the convolution theorem (Theorem A.11).

We readily see that

$$\frac{d}{d\delta}\|J_\delta f_m - f_m\|^2$$

$$= \frac{d}{d\delta}\|\hat{\rho}_\delta \hat{f}_m - \hat{f}_m\|^2$$

$$= \frac{d}{d\delta}\int_{-\infty}^{\infty}\left[\frac{1}{\sqrt{2\pi}}\exp\left(-\frac{w^2\delta^2}{4}\right)\hat{f}_m(w) - \hat{f}_m(w)\right]$$

$$\times\left[\frac{1}{\sqrt{2\pi}}\exp\left(-\frac{w^2\delta^2}{4}\right)\overline{\hat{f}_m(w)} - \overline{\hat{f}_m(w)}\right]dw$$

$$= \frac{\delta}{\sqrt{2\pi}}\int_{-\infty}^{\infty}\exp\left(-\frac{w^2\delta^2}{4}\right)\left[1 - \frac{1}{\sqrt{2\pi}}\exp\left(-\frac{w^2\delta^2}{4}\right)\right]|\hat{f}_m(w)|^2\, dw > 0.$$

$$\blacksquare$$

This shows that the mollification of the extended data function $f_m(x)$ with the Gaussian kernel $\rho_\delta(x)$ is actually an averaging process that, for a given amount of noise in the data, makes the filtered residual $\|\rho_\delta * f_m - f_m\|$ monotone with respect to δ. We have proved the following lemma.

Lemma 1.3 (Monotonicity)   Given the extended noisy data function $f_m(x)$, if $\delta_1 > \delta_2 > 0$, then

$$\|J_{\delta_1} f_m - f_m\| > \|J_{\delta_2} f_m - f_m\|.$$

The monotonicity property of Lemma 1.3 implies that there is a unique $\tilde{\delta}$ such that

$$\|J_{\tilde{\delta}} f_m - f_m\| = \varepsilon.$$

In practice, the radius of mollification $\tilde{\delta}$ is determined by solving

$$\|J_{\tilde{\delta}} f_m - f_m\|_I = \varepsilon \qquad (1.7)$$

using, for instance, the bisection method after extending the original data function f_m to the interval $I_{\tilde{\delta}}$ as explained previously.

Remarks

1. The L^2 norm represents better the global features of the function $\rho_{\delta} * f_m - f_m$ in the entire interval $I = [0, 1]$ and avoids emphasizing the behavior of $\rho_{\delta} * f_m - f_m$ near the endpoints 0 and/or 1 as might possibly be the case using the maximum norm.
2. The parameter selection (1.7) determines $\tilde{\delta}$ in a manner consistent with the amount of noise in the data function f_m. Note that if $\|f_m - f\| \leq \varepsilon$, then $\|\rho_{\delta} * f_m - f\| \leq 2\varepsilon$.

1.5   NUMERICAL PROCEDURE

To numerically approximate $d/dx[J_{\delta} f_m(x)]$, our procedure is based on the formula

$$\frac{d}{dx}[J_{\delta} f_m(x)] = (\rho_{\delta} * f)'(x),$$

a particular choice of δ, and centered differences.

Since, in practice, only a discrete set of data points is available, we assume in what follows that the data function f_m is a discrete function in $I = [0, 1]$, measured at $N + 1$ sample points $x_i = ih$, $i = 0, 1, \ldots, N$, $Nh = 1$. Given a radius of mollification, δ, we use (1.4) to extend the data to the interval $I_{\delta} = [-3\delta, 1 + 3\delta]$ and since the data are defined to be 0 in $\mathbb{R} - I_{\delta}$, we consider the extended discrete data function f_m defined at equally spaced sample points on any interval of interest.

The parameter selection is implemented by solving the discrete version of (1.7) using the bisection method. The following steps summarize the method.

Step 1. Set $\delta_{\min} = h$, $\delta_{\max} = 0.1$ and choose an initial value of δ between $\delta_{\min}$ and $\delta_{\max}$.

Step 2. Compute $J_{\delta} f_m = \rho_{\delta} * f_m$ by discrete convolution on a sufficiently large interval.

Step 3. If

$$F(\delta) = \left[\frac{1}{N+1} \sum_{i=0}^{N} \left(J_\delta f_m(x_i) - f_m(x_i) \right)^2 \right]^{1/2} = \varepsilon \pm \eta,$$

where η is a given tolerance, exit.

Step 4. If $F(\delta) - \varepsilon < -\eta$, set $\delta_{\min} = \delta$. If $F(\delta) - \varepsilon > \eta$, set $\delta_{\max} = \delta$. The updated value of δ is always given by $\frac{1}{2}(\delta_{\min} + \delta_{\max})$.

Step 5. Return to Step 2.

Once the radius of mollifcation $\tilde{\delta}$ and the discrete function $J_{\tilde{\delta}} f_m$ are determined, we use centered differences to approximate the derivative of $J_{\tilde{\delta}} f_m$ at the sample points of the interval $K_{\tilde{\delta}} = [3\tilde{\delta}, 1 - 3\tilde{\delta}]$.

1.6 NUMERICAL RESULTS

In this section we discuss the implementation of the numerical method and the tests which we have performed in order to investigate the accuracy and stability of the numerical differentiation procedure.

In all the examples, $h = 0.01$, $\delta_{\max} = 0.1$, and $I = [0, 1]$. The exact data function is denoted by $f(x)$ and the noisy data function $f_m(x)$ is obtained by adding an ε random error to $f(x)$, that is,

$$f_m(x_i) = f(x_i) + \varepsilon \theta_i, \tag{1.8}$$

where $x_i = ih$, $i = 0, 1, \ldots, N$, $Nh = 1$, and θ_i is a uniform random variable with values in $[-1, 1]$ such that

$$\max_{0 \le i \le N} |f_m(x_i) - f(x_i)| \le \varepsilon.$$

After extending the discrete data function to the interval I_δ, the parameter selection criterion was implemented with the tolerance η, used in Step 3 of the algorithm, set to reflect a 5% error in the satisfaction of the constraint. In all cases, independently of the initial choice of δ, convergence to the value $\tilde{\delta}$ determined by the selection criterion was reached in no more than eight iterations. The discrete numerical approximation to the derivative $f'(x_i)$, denoted $\tilde{f}'_m(x_i)$, is then reconstructed by means of centered differences at the sample points of $K_{\tilde{\delta}} = [3\tilde{\delta}, 1 - 3\tilde{\delta}]$. We use

$$f_{2,h} = \| f(x_i) \|_{2, K_{\tilde{\delta}}} \quad \text{and} \quad f_{\infty,h} = \| f(x_i) \|_{\infty, K_{\tilde{\delta}}}$$

to represent the l^2 norm and the maximum norm of the discrete error functions.

They are given by

$$e_{2,h} = \left\{ h \sum_{x_i \in K_{\bar{\delta}}} \left| f'(x_i) - \tilde{f}'_m(x_i) \right|^2 \right\}^{1/2}$$

and

$$e_{\infty,h} = \max_{x_i \in K_{\bar{\delta}}} \left| f(x_i) - \tilde{f}'_m(x_i) \right|,$$

respectively, according to Definitions A.13 and A.14.

EXAMPLE 1   As a first example we consider $f(x) = x(1 - \frac{1}{2}x)$. Figure 1.3 shows the solution obtained with the mollification method ($*$) and the exact

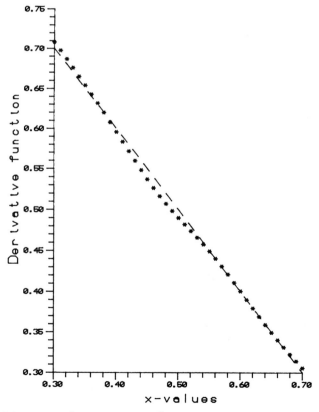

Fig. 1.3 $f(x) = x(1 - \frac{1}{2}x)$; $\varepsilon = 0.02$, $\bar{\delta} = 0.10$, $h = 0.01$. Computed derivative ($* * *$); exact (- - -).

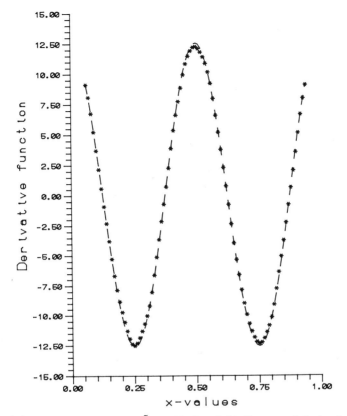

Fig. 1.4 $f(x) = \sin 4\pi x$; $\varepsilon = 0.01$, $\tilde{\delta} = 0.02$, $h = 0.01$. Computed derivative ($* * *$); exact (- - -).

derivative $f'(x) = 1 - x$. With $\varepsilon = 0.05$, the corresponding radius of molli-fication in this case is $\tilde{\delta} = 0.2$. The resolution in this problem is quite good considering the relative high noise level which we used. The associated error norms are given by $e_{2, h} = 0.001108$ and $e_{\infty, h} = 0.00576$.

EXAMPLE 2   Our second example is rather oscillatory on $[0, 1]$. We choose $f(x) = \sin 4\pi x$ and for $\varepsilon = 0.01$ we obtain $\tilde{\delta} = 0.02$. In Fig. 1.4 we plot the numerical solution obtained with the mollification method ($*$) and the exact derivative $f'(x) = 4\pi \cos 4\pi x$. The corresponding relative error norms are given by

$$\frac{e_{2, h}}{\left(\tilde{f}'_m\right)_{2, h}} = 0.01976 \quad \text{and} \quad \frac{e_{\infty, h}}{\left(\tilde{f}'_m\right)_{\infty, h}} = 0.02439.$$

EXAMPLE 3   In Fig. 1.5 we plot the exact derivative of $f(x) = \sin 10\pi x$ and the solution obtained with the mollification method ($*$). The data

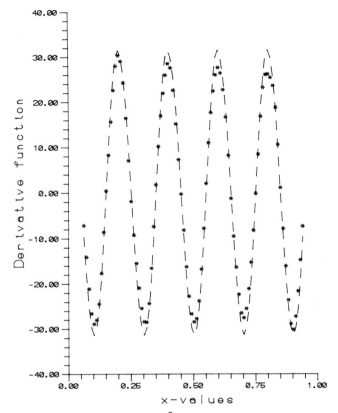

Fig. 1.5 $f(x) = \sin 10\pi x$; $\varepsilon = 0.10$, $\tilde{\delta} = 0.02$, $h = 0.01$. Computed derivative ($* * *$); exact (- - -).

function is highly oscillatory and the noise level $\varepsilon = 0.1$ very high. The resulting radius of mollification is $\tilde{\delta} = 0.02$. The relative error norms are given by $e_{2,h}/(\tilde{f}'_m)_{2,h} = 0.11180$ and $e_{\infty,h}/(\tilde{f}'_m)_{\infty,h} = 0.13282$. We conclude that even in this rather difficult case, the method performs quite satisfactorily.

Finally, in order to investigate the stability of the numerical method, we would like to determine the amplification factor associated with errors in the data when using the numerical differentiation procedure. If we set $f(x_i) = 0$ in (1.8), we can compute the response of the method to pure noise as a function of the radius of mollification and thereby get a measure of the amplification factor. Since all the response norms are essentially proportional, we only plot in Fig. 1.6 a representative curve for $\varepsilon = 0.1$. The solid curve gives $\| \cdot \|_{\infty, I}$ and the dashed curve corresponds to $\| \cdot \|_{2, I}$.

We notice that the "derivative of the noise" has been computed on $I = [0, 1]$ for every value of δ. To obtain these responses, we read the discrete

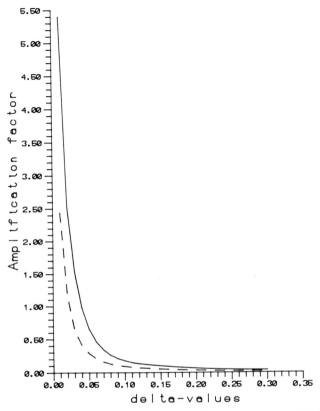

Fig. 1.6 Response to noise as a function of δ. $f_m(x_i) = \varepsilon\theta_i$, $\varepsilon = 0.1$. $\|\cdot\|_{\infty, I}$ (—); $\|\cdot\|_{2, I}$ (- - -).

data on the interval $[-1, 2]$ with $h = 0.01$ and performed the discrete convolution in $[-0.5, 1.5]$ for each value of δ. Then we applied centered differences to obtain the approximate derivative in $I = [0, 1]$ using the corresponding δ value.

1.7 EXERCISES

1.1. For $\delta \geq h > 0$, k integer, consider the grid points $x = kh$ in the whole discrete line and define the discrete—see Sec. A.4—averaging kernel $\rho_{k, \delta}$ by

$$\rho_{h, \delta} = \begin{cases} \dfrac{1}{C_\delta} \exp\left[\dfrac{k^2 h^2}{k^2 h^2 - \delta^2} \right], & |kh| < \delta, \\ 0, & |kh| \geq \delta, \end{cases}$$

where $C_\delta = h\sum_{-N}^{N} \exp[k^2 h^2/(k^2 h^2 - \delta^2)]$. Here N denotes the largest integer less than or equal to δ/h. Prove that

(a) $h\sum_k \rho_{h,\delta} = 1$.

(b) Given a discrete function g_h,

$$(D_0(\rho_{h,\delta} * g_h))(kh) = (\rho_{h,\delta} * (D_0 g)_h)(kh),$$

where $(f_h * g_h)(kh) = h\sum_j f(jh)g((k-j)h)$.

1.2. (*Discrete analogous of Lemma* 1.1) With $\rho_{h,\delta}$ as in Exercise 1.1, if $\|D_+^2 g_h\|_\infty < M$, show that

$$\|D_0(\rho_{h,\delta} * g_h) - D_0 g_h\|_\infty \le \delta M$$

and

$$\|D_+(\rho_{h,\delta} * g_h) - D_+ g_h\|_\infty \le \delta M.$$

1.3. With $\rho_{h,\delta}$ as in Exercise 1.1, show that if δ is a multiple of h, then

$$D_0(\rho_{h,\delta} * g_h) = (D_0 \rho_{h,\delta}) * g_h.$$

1.4. (*Discrete analogous of Lemma* 1.2) With $\rho_{h,\delta}$ as in Exercise 1.1, if δ is a multiple of h and $g_{h,m}$ is a noisy discrete function such that $\|g_h - g_{h,m}\|_\infty \le \varepsilon$, show that

$$\|D_0(\rho_{h,\delta} * g_{h,m}) - D_0(\rho_{h,\delta} * g_h)\|_\infty \le \frac{2\varepsilon}{3\delta}.$$

1.5. (*Discrete error estimate*) Under the conditions of Exercises 1.2 and 1.4, show that

$$\|D_0(\rho_{h,\delta} * g_{h,m}) - D_0 g_h\|_\infty \le \delta M + \frac{2\varepsilon}{3\delta}.$$

1.6. Show that if δ is a multiple of h,

$$\|g_m - g\|_\infty \le \varepsilon \qquad \text{and} \qquad \sup_{x \in \mathbb{R}} \left|\frac{d^i}{dx^i} g(x)\right| \le M, \qquad i = 2, 3,$$

under the conditions of Exercises 1.2 and 1.4,

$$\left\|D_0(\rho_{h,\delta} * g_{h,m}) - \left(\frac{d}{dx}g\right)_h\right\|_\infty \le (1+h)\delta M + \frac{2\varepsilon}{3\delta} + \frac{h^2 M}{6}.$$

Here $(d/dx\, g)_h$ indicates the restriction of g' to the grid points.

1.7. Consider the heat conduction problem in a material that is undergoing radioactive decay or damage. We pose the identification problem of attempting to determine a positive function $a(t)$ and temperature function $u(x, t)$ that satisfy

$$u_t = a(t)u_{xx}, \qquad 0 < x < 1, 0 < t,$$

$$u(0, t) = u(1, t) = 0, \qquad 0 < t,$$

$$u(x, 0) = \sin \pi x, \qquad 0 < x < 1,$$

$$u(1/2, t) = g(t), \qquad 0 < t, g > 0, g(0) = 1,$$

with approximate data function $g_m(t)$, $\|g - g_m\|_{2, [0, 1]} \le \varepsilon$. The function $g(t)$, a measured interior data temperature, constitutes the necessary data over specification for the unique determination of the function $a(t)$. Show that

(a)
$$u(x, t) = \exp\left[-\pi^2 \int_0^t a(y)\, dy\right] \sin \pi x,$$

(b)
$$g(t) = \exp\left[-\pi^2 \int_0^t a(y)\, dy\right],$$

(c)
$$a(t) = -\frac{g'(t)}{\pi^2 g(t)}, \qquad t > 0.$$

1.8. Solve Exercise 1.7 numerically, using the mollification method, if $g(t) = \exp[-\pi^2 t^2/2]$, $0 \le t \le 1$, with $h = \Delta t = 0.01$ after generating the discrete noisy data $g_{h, m}(kh) = g(kh) + \varepsilon_k$ by adding an ε_k random noise to the restriction of the exact function g to the grid points. Here the ε_k's are independent Gaussian random variables with variance $\sigma = \varepsilon^2$. Use $\varepsilon = 0.000$, 0.001, and 0.010. Measure the discrete l^2 error norm using the sample points of the interval $K_{\tilde\delta} = [3\tilde\delta, 1 - 3\tilde\delta]$ after verifying that the exact solution is given by $a(t) = t$, $0 \le t \le 1$.

1.8   REFERENCES AND COMMENTS

The following references and comments serve to expand the basic material covered in the corresponding sections.

1.1. Numerical differentiation has been discussed by many authors and a number of solution methods have been proposed. Finite-difference approaches have been used, for example, by

L. W. Johnson and R. D. Riess, An error analysis for numerical differentiation, *J. Inst. Math. Appl.* **11** (1973), 115–120.

J. N. Lyness, Has numerical differentiation a future?, *Proc. Seventh Manitoba Conf. on Numerical Mathematics and Computing, Congress, Numer., Utilitas, Math.,* Winnipeg, Manitoba, 1978, pp. 107–129.

J. Oliver, An algorithm for numerical differentiation of a function of one real variable, *J. Comput. Appl. Math.* **6** (1980), 145–160.

T. Strom and J. N. Lyness, On numerical differentiation, *BIT* **15** (1975), 314–322.

The use of Tikhonov's regularization technique applied to the problem of numerical differentiation has been thoroughly treated by

J. Cullum, Numerical differentiation and regularization, *SIAM J. Numer. Anal.* **8** (1975), 254–265.

N. S. Surova, An investigation of the problem of reconstructing a derivative by using an optimal regularizing integral operator, *Numer. Methods Programming* **1** (1977), 30–34.

The stability of the numerical process by means of finite-dimensional regularization, using finite elements, can be found in the more recent work by

J. T. King and D. A. Murio, Numerical differentiation by finite-dimensional regularization, *IMA J. Numer. Anal.* **6** (1986), 65–85.

For methods of a statistical nature, excellent references are

K. S. Anderssen and P. Bloomfield, A time series approach to numerical differentiation, *Technometrics* **16** (1974), 69–75.

M. L. Baart, Computational experience with the spectral smoothing method for differentiating noisy data, *J. Comput. Phys.* **42** (1981), 141–151.

D. D. Cox, Asymptotics of *M*-type smoothing splines, *Ann. Statist.* **11** (1983), 530–551.

J. Rice and M. Rosenblatt, Smoothing splines: regression, derivatives, and disconvolution, *Ann. Statist.* **11** (1983), 141–156.

Historically, the distinction between the concepts of well-posed problems and ill-posed problems has been initially emphasized in the classical work of

J. Hadamard, *Le Problème de Cauchy*, Herman et Cie., Paris, 1932.

1.2. The basic idea of attempting to reconstruct a mollified version of the unknown function utilizing a quadrature algorithm based on the differentiation of the convolution kernel was first reported in

V. V. Vasin, The stable evaluation of a derivative in space $C(-\infty, \infty)$, *U.S.S.R. Computational Math. and Math. Phys.* **13** (1973), 16–24.

For a totally different problem, the introduction of the Gaussian kernel to perform the mollification and greatly facilitate the associated Fourier analysis was originally illustrated in

P. Manselli and K. Miller, Calculation of the surface temperature and heat flux on one side of a wall from measurements on the opposite side, *Ann. Mat. Pura Appl.* (4) **123** (1980), 161–183.

The mollification method as presented in this section, generated by initially filtering the noisy data by discrete convolution against a suitable averaging kernel and then using finite centered differences to numerically compute the corresponding well-posed problem, follows very closely

D. A. Murio, Automatic numerical differentiation by discrete mollification, *Comput. Math. Appl.* **13** (1987), 381–386.

1.3–1.6. First-kind Fredholm and Volterra integral equations have always been considered as prototypes for ill-posed inverse problems. Some of the recommended books on Fredholm and Volterra integral equations which cover basic as well as general aspects of the theory while stressing also the numerical aspects are

H. Bruner and P. J. van der Houwen, *The Numerical Solution of Volterra Equations*, North-Holland, Amsterdam, 1986.

P. Linz, *Analytical and Numerical Methods for Volterra Equations*, SIAM, Philadelphia, 1985.

A finite-element asymptotic analysis of the finite-dimensional regularization method of Tikhonov, in the context of Hilbert spaces, for first-kind Fredholm integral equations, can be found in

C. W. Groetsch, J. T. King, and D. A. Murio, Asymptotic analysis of a finite element method for Fredholm equations of the first kind, *Treatment of Integral Equations by Numerical Methods*, C. T. H. Baker and G. F. Miller (eds.), Academic, London, 1984, pp. 1–11.

For an extended treatment of the theoretical aspects of Tikhonov's regularization applied to first-kind Fredholm equations, from the classical point of view of compact operators in Hilbert spaces with unbounded inverses, see

C. W. Groetsch, *The Theory of Tikhonov Regularization for Fredholm Equations of the First Kind*, Pitman, London, 1984.

1.7. The totally discretized problem has been treated in detail in

D. A. Murio and L. Guo, Discrete stability analysis of the mollification method for numerical differentiation, *Comput. Math. Appl.* **19** (1990), 15–26.

Exercise 1.7 is taken from

J. R. Cannon, *The One-Dimensional Heat Equation*, Addison-Wesley, Reading, MA, 1984,

where the problem is considered under the assumption that the data are known exactly.

2

ABEL'S INTEGRAL EQUATION

The difficult problem of determining the structure of an object from its three-dimensional cone-beam data projections, in computerized tomography, is currently receiving considerable attention. When the object is known to be radially symmetric, its structure can be determined by using the inverse Abel transform. If the object does not have radial symmetry, it can be reconstructed, in principle, by using the inverse Radon transform.

2.1 DESCRIPTION OF THE PROBLEM

To state the problem of reconstruction formally, consider a circular cross section of radius 1, perpendicular to the axis of symmetry. If the direction of radiation is parallel to the y axis, the radiation intensity $I(y)$ and the variable absorption coefficient $a(x, y)$—defined over the unit circle—are related by the differential equation $dI/dy = -aI$. Thus, assuming that the absorption of radiation outside the circle is negligible—the geometry is illustrated in Fig. 2.1—we have

$$I_x = I_0 \exp\left(-\int_{-\sqrt{1-x^2}}^{\sqrt{1-x^2}} a(x, y)\, dy\right).$$

Introducing $F(x) = \ln(I_0/I_x)$, the preceding formula becomes

$$F(x) = \int_{-\sqrt{1-x^2}}^{\sqrt{1-x^2}} a(x, y)\, dy.$$

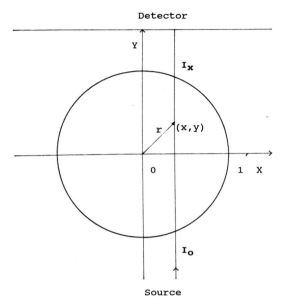

Fig. 2.1   Intensity of radiation I_x on the unit circle parallel to the y axis.

If the absorption coefficient is radially distributed, that is, if $a(x, y) = a(r)$, $r^2 = x^2 + y^2$, we obtain instead, the singular integral

$$F(x) = \int_x^1 \frac{2r}{\sqrt{r^2 - x^2}} a(r) \, dr. \tag{2.1}$$

Making the successive change of variables $s = 1 - r^2$ and $t = 1 - x^2$, (2.1) can be written as

$$F(\sqrt{1 - t}) = \int_0^t \frac{a(\sqrt{1 - s})}{\sqrt{t - s}} \, ds.$$

It is now natural to define the new functions g and f by $g(s) = a(\sqrt{1 - s})$ and $f(t) = F(\sqrt{1 - t})$, to get

$$f(t) = \int_0^t \frac{g(s)}{\sqrt{t - s}} \, ds, \qquad 0 < t \le 1. \tag{2.2}$$

This is a "classical" Abel type of integral equation and its inversion and numerical solution have been widely studied. We want to determine the "structure of the object," g—in this case an absorption coefficient—from knowledge of its projections along a direction, f.

In order to invert Abel's equation, we multiply both sides of (2.2) by the kernel $(z - t)^{-1/2}$ and integrate with respect to t. We get

$$\int_0^z \frac{f(t)}{\sqrt{z - t}} \, dt = \int_0^z \frac{1}{\sqrt{z - t}} \left(\int_0^t \frac{g(s)}{\sqrt{t - s}} \, ds \right) dt$$

$$= \int_0^z g(s) \left(\int_s^z \frac{dt}{\sqrt{(z - t)(t - s)}} \right) ds$$

$$= \int_0^z g(s) \left(\int_0^{z-s} \frac{du}{\sqrt{u(x - s - u)}} \right) ds$$

$$= 2 \int_0^z g(s) \left(\int_0^1 \frac{dv}{\sqrt{1 - v^2}} \right) ds$$

$$= \pi \int_0^z g(s) \, ds,$$

after interchanging the order of integration and introducing the change of variables $u = t - s$ and $v = (u/(t - s))^{1/2}$ in the inner integral. Finally, differentiating with respect to z, we obtain

$$g(z) = \frac{1}{\pi} \frac{d}{dz} \int_0^z \frac{f(t)}{\sqrt{z - t}} \, dt, \qquad 0 < t \le 1. \qquad (2.3) \quad \blacksquare$$

Notice that in the preceding proof, we were assuming the integrability of $f(t)/\sqrt{t}$. This is certainly the case if $f(t)$ is continuous in $(0, 1]$ and as t approaches 0, the integrand is allowed to grow not faster than t^{-q}, $q < 1$. (Recall that t^{-1} is not integrable in a neighborhood of 0.) Also notice that the solution function $g(t)$ is continuous in $(0, 1]$. Thus, we have proved the following theorem.

Theorem 2.1 (Inverse Abel's Transform)   If $f(t)$ is continuous in $0 < t \le 1$ and $\lim_{t \to 0} t^\beta f(t) = c$, $c \ne 0$, and $\beta < 1/2$, then (2.2) has the unique solution

$$g(t) = \frac{1}{\pi} \frac{d}{dt} \int_0^t \frac{f(s)}{\sqrt{t - s}} \, ds, \qquad 0 < t \le 1.$$

We can go one step further by attempting to perform the differentiation but because of the singularity in the kernel, this cannot be carried out explicitly.

However, if we integrate by parts, assuming that $g(t)$ has a continuous first derivative in $(0, 1)$, we get

$$
g(t) = \frac{1}{\pi} \frac{d}{dt} \left[-2f(s)\sqrt{t-s} \Big|_0^t + 2\int_0^t f'(s)\sqrt{t-s}\, ds \right]
$$

$$
= \frac{1}{\pi} \frac{d}{dt} \left[2f(0)\sqrt{t} + 2\int_0^t f'(s)\sqrt{t-s}\, ds \right]
$$

$$
= \frac{1}{\pi} \left[\frac{f(0)}{\sqrt{t}} + \int_0^t \frac{f'(s)}{\sqrt{t-s}}\, ds \right], \qquad 0 < t \le 1.
$$

In what follows we assume f to be continuously differentiable and, without loss of generality, we consider $f(0) = 0$. [Note that if this is not the case, we can always subtract $f(0)$ from our original data function.] Hence, if the original data function f is given by (2.2), the solution function g is given by

$$
g(t) = \frac{1}{\pi} \int_0^t \frac{f'(s)}{\sqrt{t-s}}\, ds, \qquad 0 < t \le 1. \tag{2.4}
$$

As we have already seen in Chap. 1, if f is obtained through measured data, small errors might cause large errors in the computation of the derivative f' which in turn is needed to evaluate the solution function g. The inverse problem is ill conditioned and, consequently, the direct use of the Volterra integral equation of the first kind (2.4) is *very limited* and special methods are needed.

How ill posed is this problem? At this point, we can only attempt to compare the inversion of Abel's integral equation with the process of numerical differentiation.

In order to perform some frequency analysis, we extend the data function f and the kernel function $k(t) = 1/\sqrt{t}$ to be 0 for $t < 0$ and consider their corresponding Fourier transforms. With $\hat{k}(w) = (1 - i)/2\sqrt{w}$, $i = \sqrt{-1}$, we have

$$
f(t) = \int_{-\infty}^{\infty} \frac{g(s)}{\sqrt{t-s}}\, ds = \int_{-\infty}^{\infty} k(t-s)g(s)\, ds = (k * g)(t), \qquad -\infty < t < \infty.
$$

Thus,

$$
\hat{f}(w) = \hat{k}(w) \cdot \hat{g}(w) = \frac{1-i}{2\sqrt{w}} \hat{g}(w) \tag{2.5}
$$

and

$$\hat{g}(w) = (1 + i)\sqrt{w}\,\hat{f}(w) = \sqrt{w}\,\hat{f}(w) + i\frac{w}{\sqrt{w}}\hat{f}(w)$$

$$= \sqrt{w}\,\hat{f}(w) + \frac{\hat{f}'(w)}{\sqrt{w}}$$

$$= \sqrt{w}\,\hat{k}(w) \cdot \hat{g}(w) + \frac{1}{\sqrt{w}}\hat{k}(w) \cdot \hat{g}'(w)$$

$$= \frac{1}{i\sqrt{w}}\hat{k}(w) \cdot \hat{g}'(w) + \frac{1}{\sqrt{w}}\hat{k}(w) \cdot \hat{g}'(w), \qquad (2.6)$$

after using (2.5) and the property $\hat{f}'(w) = \hat{k}(w) \cdot \hat{g}'(w)$.
   Consequently,

$$\hat{g}(w) = \frac{1}{\sqrt{w}}(1 - i)\hat{k}(w) \cdot \hat{g}'(w) = 2\hat{k}(w) \cdot \hat{k}(w) \cdot \hat{g}'(w)$$

and

$$\hat{g}(w) = \left(\sqrt{2}\,\hat{k}(w)\right)\left(\sqrt{2}\,\hat{k}(w)\right)\hat{g}'(w). \qquad (2.7) \quad \blacksquare$$

   This can be interpreted as saying that the application of $\sqrt{2}\,\hat{k}$, twice, eliminates the differentiation. This explains the "half differentiation" characterization of Abel's inversion process.
   For the differentiation process, in frequency space,

$$\hat{f}'(w) = iw\hat{f}(w), \qquad f \text{ data, } f' \text{ unknown,} \qquad (2.8)$$

and, by linearity, the high-frequency components of the error are also amplified by the factor $|w|$.
   For Abel's problem, from (2.6), we observe that

$$\hat{g}(w) = (1 + i)\sqrt{w}\,\hat{f}(w), \qquad f \text{ data, } g \text{ unknown,} \qquad (2.9)$$

and, once again, the cause of the ill-posedness is due to the presence of high-frequency components in the data error. This time they are amplified by the factor $\sqrt{2|w|}$.
   We safely conclude that Abel's problem is less ill posed than the differentiation problem.

2.2 STABILIZED PROBLEMS

In this section we introduce and analyze several special methods that can be used to stabilize the inverse Abel problem under the assumption that the measured data function f_m satisfies the error bound $\|f - f_m\|_{\infty, I} \leq \varepsilon$ in the data interval $I = [0, 1]$. Their actual numerical implementations will be discussed in the next section.

We begin with the mollification method—Method I here—as presented in Chap. 1.

Method I

After introducing the δ-mollifier

$$\rho_\delta(t) = \frac{1}{\delta \sqrt{\pi}} \exp\left[\frac{-t^2}{\delta^2}\right]$$

and extending the data functions f and f_m to the interval $I_\delta = [-3\delta, 1 + 3\delta]$ in such a way that they decay smoothly to 0 in $[1, 1 + 3\delta]$ and they are 0 in $\mathbb{R} - [0, 1 + 3\delta]$, we define the following functions taking into account (1.3):

$$g_\delta(t) = \frac{1}{\pi} \int_0^t (J_\delta f)'(s)(t - s)^{-1/2} ds, \qquad 0 < t \leq 1, \qquad (2.10)$$

and

$$g_\delta^\varepsilon(t) = \frac{1}{\pi} \int_0^t (J_\delta f_m)'(s)(t - s)^{-1/2} ds, \qquad 0 < t \leq 1. \qquad (2.11)$$

The following two lemmas give the necessary estimates.

Lemma 2.1 (Consistency)   Under the conditions of Lemma 1.1,

$$\|g - g_\delta\|_{\infty, I} \leq \frac{6}{\pi} \delta M_2.$$

Proof. From (2.4) and (2.10), for $t \in I$,

$$|g(t) - g_\delta(t)| \leq \frac{1}{\pi} \int_0^t |f'(s) - (J_\delta f_m)'(s)|(t - s)^{-1/2} ds$$

$$\leq \frac{1}{\pi} \|f' - (J_\delta f_m)'\|_{\infty, I} \int_0^t (t - s)^{-1/2} ds$$

$$\leq \frac{6}{\pi} \delta M_2 \qquad \text{by Lemma 1.1.} \quad \blacksquare$$

Lemma 2.2 (Stability)   Under the conditions of Lemma 1.2,

$$\|g_\delta^\varepsilon - g_\delta\|_{\infty, I} \le \frac{4}{\pi^{3/2}} \frac{\varepsilon}{\delta}.$$

Proof. From (2.10) and (2.11), for $t \in I$,

$$|g_\delta(t) - g_\delta^\varepsilon(t)| \le \frac{1}{\pi} \int_0^t |(J_\delta f)'(s) - (J_\delta f_m)'(s)|(t - s)^{-1/2} \, ds$$

$$\le \frac{1}{\pi} \|(J_\delta f)' - (J_\delta f_m)'\|_{\infty, I} \int_0^t (t - s)^{-1/2} \, ds$$

$$\le \frac{4}{\pi^{3/2}} \frac{\varepsilon}{\delta} \qquad \text{by Lemma 1.2.} \quad \blacksquare$$

Lemma 2.2 shows that attempting to reconstruct the approximate inverse Abel transform function g_δ is a stable problem with respect to perturbations in the data function, in the maximum norm, and for fixed δ.

The usual error estimate follows:

Theorem 2.2 (Error Estimate)   Under the conditions of Lemmas 2.1 and 2.2,

$$\|g_\delta^\varepsilon - g\|_{\infty, I} \le \frac{2}{\pi} \left[3\delta M_2 + \frac{2}{\sqrt{\pi}} \frac{\varepsilon}{\delta} \right]. \tag{2.12}$$

Proof. The estimate follows from Lemmas 2.1 and 2.2 and the triangle inequality. $\blacksquare$

As was also the case with numerical differentiation, the error estimate is minimized by choosing $\delta = \bar{\delta} = [2\varepsilon/3\sqrt{\pi} M_2]^{1/2}$ and we obtain uniform convergence as $\varepsilon \to 0$ with rate $O(\varepsilon^{1/2})$. This "optimal" selection of the radius of mollification is invalidated in practice because M_2 is not known, in general. However, we can always use the parameter selection criterion discussed in Sec. 1.4.

The complete abstract algorithm is as follows:

Step 1. Smoothly extend the noisy data function f_m to the interval $I_{\delta_{\max}} = [-3\delta_{\max}, 1 + 3\delta_{\max}]$.

Step 2. Automatically determine the unique radius of mollification δ as a function of the level of noise ε.

Step 3. Compute the filtered data function $J_\delta f_m$.

Step 4. Evaluate g_δ^ε using (2.11).

Next, we introduce Method II, which is based on the mollification method as originally proposed for two-dimensional reconstruction geometries.

Method II

First, we notice that the exact inverse formula (2.4)—with the corresponding extension of the functions involved—can be written as the convolution equation

$$g(t) = \frac{1}{\pi}(k * f')(t), \qquad 0 < t \leq 1.$$

The mollification of this convolution equation with the Gaussian kernel ρ_μ gives the new formula $(\rho_\mu * g)(t) = 1/\pi(\rho_\mu *(k * f'))(t), 0 < t \leq 1$.

In Method I we associated the right-hand side as $1/\pi k *(\rho_\mu * f)'$; for Method II we associate, instead, as $1/\pi(\rho_\mu * k)' * f$ and obtain the approximate reconstruction solution

$$g_\mu^\varepsilon(t) = \frac{1}{\pi}\int_0^t (J_\mu k)'(t - s)f_m(s)\,ds, \qquad 0 < t \leq 1. \tag{2.13}$$

Mathematically, formula (2.11), for Method I, and formula (2.13), for Method II, are identical. Consequently, the theoretical error bound (2.12)—derived for Method I—also applies to Method II. However, the implementation is now quite different since the Gaussian mollification of the kernel function in the interval $I = [0, 1]$ requires the extension of $k(t)$ to the interval $I_\mu = [-3\mu, 1 + 3\mu]$. We accomplish this by defining $k(t) = k(-t)$ for $t < 0$. Notice that in Method II the function f_m need not be extended outside the unit interval.

The complete abstract algorithm for Method II is as follows:

Step 1. Extend the kernel function k as an even function of t for $t < 0$.
Step 2. Choose $\mu > 0$.
Step 3. Compute the mollified kernel $J_\mu k$.
Step 4. Evaluate g_μ^ε using (2.13).

Remarks

1. In Method II the mollified kernel is computed only once and can be used repeatedly for many different data functions.
2. Method I requires filtering each data function one at a time. The corresponding radii of mollification can be automatically selected according to the quality of the measured data.

3. The selection of the mollification parameter in Method II requires further considerations.

4. In both cases, the actual implementation of the algorithms requires the choice of a suitable quadrature formula to approximate the integrals in (2.11) and (2.13).

Method III

This method is based on a reconstruction technique motivated by the singularity of the kernel. The algorithm was initially proposed for arbitrary two-dimensional ray schemes in computerized tomography and more recently has been extended to three-dimensional image reconstruction methods from cone-beam projections.

From sufficiently smooth errorless data, after integrating by parts—both ways—in (2.4), we obtain the equivalent solution expressions

$$g(t) = \frac{1}{\pi}\left\{2\sqrt{t}\,f'(0) + 2\int_0^t f''(s)\sqrt{t-s}\,ds\right\}, \qquad 0 \le t \le 1, \quad (2.14)$$

and

$$g(t) = \frac{1}{\pi}\lim_{\gamma \to 0}\left\{\frac{1}{\sqrt{\gamma}}f(t-\gamma) - \frac{1}{2}\int_0^{t-\gamma} f(s)(t-s)^{-3/2}\,ds\right\}, \qquad 0 \le t \le 1.$$

Using Taylor's expansions, we observe that

$$\gamma^{-1/2}f(t-\gamma) - \gamma^{-3/2}\int_{t-\gamma}^t f(s)\,ds = -\frac{1}{2}\gamma^{1/2}f'(t) + O(\gamma^{3/2}) \quad (2.15)$$

and, consequently,

$$\frac{1}{\pi}\lim_{\gamma \to 0}\left\{\frac{1}{\sqrt{\gamma}}f(t-\gamma) - \gamma^{-3/2}\int_{t-\gamma}^t f(s)\,ds\right\} = 0. \qquad (2.16)$$

Subtracting (2.16) from (2.14) gives

$$g(t) = \frac{1}{\pi}\lim_{\gamma \to 0}\left\{\gamma^{-3/2}\int_{t-\gamma}^t f(s)\,ds - \frac{1}{2}\int_0^{t-\gamma} f(s)(t-s)^{-3/2}\,ds\right\},$$

$$0 \le t \le 1.$$

The approximate inverse Abel transform is now obtained by eliminating the limit procedure in the last expression, that is,

$$g_\gamma(t) = \frac{1}{\pi} \left\{ \gamma^{-3/2} \int_{t-\gamma}^{t} f(s)\, ds - \frac{1}{2} \int_{0}^{t-\gamma} f(s)(t-s)^{-3/2}\, ds \right\}, \qquad 0 \le t \le 1.$$

$$(2.17)$$

Notice that by defining the kernel function $(0 < \gamma < 1)$

$$H_\gamma(t) = \begin{cases} \gamma^{-3/2}, & 0 \le t < \gamma, \\ -\dfrac{1}{2} t^{-3/2}, & \gamma \le t, \end{cases} \qquad (2.18)$$

the approximate solution can also be written as

$$g_\gamma(t) = \frac{1}{\pi} (H_\gamma * f)(t), \qquad 0 \le t \le 1. \qquad (2.19)$$

The family of kernel functions $H_\gamma(t)$ satisfies the normalizing property

$$\int_0^1 H_\gamma(t)\, dt = 1, \qquad 0 < \gamma < 1.$$

The usual two lemmas for Method III—the regularization by truncation method—are given below.

Lemma 2.3 (Consistency)   If $f(0) = 0$, $\|f'\|_{\infty, I} \le M_1$, and $f'' \in C^0(I)$, then

$$\|g - g_\gamma\|_{\infty, I} \le \frac{5}{2\pi} \sqrt{\gamma}\, M_1 + O(\gamma^{3/2}).$$

Proof. Integrating by parts in (2.17) and using (2.15) plus the fact that $f(0) = 0$, we obtain

$$g_\gamma(t) = \frac{1}{\pi} \left\{ \frac{1}{2} \sqrt{\gamma} f'(t) + O(\gamma^{3/2}) + \int_0^{t-\gamma} f'(s)(t-s)^{-1/2}\, ds \right\},$$

$$0 \le t \le 1.$$

Integrating by parts again, recalling that $f'' \in C(I)$,

$$g_\gamma(t) = \frac{1}{\pi} \left\{ \frac{1}{2} \sqrt{\gamma} f'(t) - \frac{2}{\pi} f'(t-\gamma) + \frac{2}{\pi} f'(0)\sqrt{t} \right.$$

$$\left. + 2 \int_0^{t-\gamma} f''(s)\sqrt{t-s}\, ds + O(\gamma^{3/2}) \right\}, \qquad 0 \le t \le 1.$$

A comparison of this formula with expression (2.14) allows us to write

$$g_\gamma(t) - g(t) = \frac{1}{\pi}\left\{\sqrt{\gamma}\left(\frac{1}{2}f'(t) - 2f'(t-\gamma)\right)\right.$$

$$\left. -2\int_{t-\gamma}^t f''(s)\sqrt{t-s}\,ds\right\} + O(\gamma^{3/2}), \qquad 0 \le t \le 1.$$

Observing that

$$\left|\frac{2}{\pi}\int_{t-\gamma}^t f''(s)\sqrt{t-s}\,ds\right| \le \frac{4}{3\pi}\gamma^{3/2}\|f''\|_{\infty,I}$$

and

$$\left|\frac{1}{2}f'(t) - 2f'(t-\gamma)\right| \le \frac{1}{2}|f'(t)| + 2\gamma\|f''\|_{\infty,I},$$

it follows that

$$|g_\gamma(t) - g(t)| \le \frac{5}{2\pi}\sqrt{\gamma}|f'(t)| + O(\gamma^{3/2}), \qquad 0 \le t \le 1.$$

Thus,

$$\|g - g_\gamma\|_{\infty,I} \le \frac{5}{2\pi}\sqrt{\gamma}M_1 + O(\gamma^{3/2}). \qquad \blacksquare$$

Lemma 2.4 (Stability)   If $f_m \in C^0(I)$ and $\|f - f_m\|_{\infty,I} \le \varepsilon$, then

$$\|g_\gamma - g_\gamma^\varepsilon\|_{\infty,I} \le \frac{2}{\pi}\frac{\varepsilon}{\sqrt{\gamma}}.$$

Proof. Using (2.17), we get

$$g_\gamma^\varepsilon(t) - g_\gamma(t) = \frac{1}{\pi}\left\{\gamma^{-3/2}\int_{t-\gamma}^t (f_m(s) - f(s))\,ds\right.$$

$$\left. -\frac{1}{2}\int_0^{t-\gamma}(f_m(s) - f(s))(t-s)^{-3/2}\,ds\right\},$$

$$0 \le t \le 1.$$

Hence,

$$\left| g_\gamma^\varepsilon - g_\gamma(t) \right| \le \frac{\varepsilon}{\pi} \left\{ \gamma^{-1/2} + \left(\gamma^{-1/2} - t^{-1/2} \right) \right\} \le \frac{2}{\pi} \varepsilon \gamma^{-1/2}, \qquad 0 \le t \le 1,$$

and

$$\| g_\gamma - g_\gamma^\varepsilon \|_{\infty, I} \le \frac{2}{\pi} \frac{\varepsilon}{\sqrt{\gamma}}. \quad \blacksquare$$

Lemma 2.4 shows that attempting to reconstruct the approximate inverse Abel transform function g_γ is a stable problem with respect to perturbations in the data function, in the maximum norm, and for γ fixed.

As usual, we also have for Method III, the following error estimate.

Theorem 2.3 (Error Estimate)   Under the conditions of Lemmas 2.3 and 2.4,

$$\| g_\gamma^\varepsilon - g \|_{\infty, I} \le \frac{5}{2\pi} \sqrt{\gamma} M_1 + \frac{2}{\pi} \frac{\varepsilon}{\sqrt{\gamma}} + O(\gamma^{3/2}).$$

The "optimal" choice of the regularizing parameter $\gamma = \bar{\gamma} = (4/5)\varepsilon/M_1$—minimizing the upper bound of the error estimate—gives uniform convergence as $\varepsilon \to 0$ with rate $O(\varepsilon^{1/2})$.

Method IV

This method is the well-known regularization procedure of Tikhonov. In a more abstract setting, (2.2) can be written as

$$Ag = f, \tag{2.20}$$

where A is the "data operator"—Abel's operator in our case—and we assume that the unknown function g and the data function f belong to some suitable subset of smooth L^2 functions on the interval $I = [0, 1]$.

It is quite natural to assume that the unknown function g satisfies an L^2 error bound of the form

$$\| Ag - f \|_{2, I}^2 = (Ag - f, Ag - f) = \int_0^1 [Ag(t) - f(t)]^2 \, dt \le \varepsilon. \tag{2.21}$$

In order to help stabilize the inverse problem, we hypothesize that the unknown function g itself satisfies an a priori global L^2 bound

$$\| g \|_{2, I}^2 = (g, g) = \int_0^1 [g(t)]^2 \, dt \le E^2. \tag{2.22}$$

We notice that by replacing g with some higher-order derivative of g in (2.22), it is possible to impose a much stronger condition on the unknown function g.

If g satisfies (2.21) and (2.22), it also satisfies

$$\frac{1}{2}\left(\|Ag - f\|_{2,I}^2 + \left(\frac{\varepsilon}{E}\right)^2 \|g\|_{2,I}^2 \right) \leq \varepsilon^2.$$

We can now choose our approximation for the function g as the one minimizing the functional

$$S_\alpha(g) = \frac{1}{2}\left(\|Ag - f\|_{2,I}^2 + \alpha\|g\|_{2,I}^2 \right). \tag{2.23}$$

Here, $\alpha = (\varepsilon/E)^2$ is the regularization parameter—a Lagrange multiplier—with a clear physical meaning.

In what follows we will indicate the unique solution for the minimization problem (2.23) by g_α and $\|\cdot\|_{2,I}$ by $\|\cdot\|$.

The set of linear canonical equations associated with the minimization of the functional (2.23) can be obtained by considering an arbitrary real number λ and an arbitrary function g satisfying (2.22).

In fact,

$$2S_\alpha(g_\alpha + \lambda g) = \|A(g_\alpha + \lambda g) - f\|^2 + \alpha\|g_\alpha + \lambda g\|^2$$

$$= S_\alpha(g_\alpha) + 2\lambda[(Ag_\alpha - f, Ag) + \alpha(g, g_\alpha)]$$

$$+ \lambda^2[(Ag, Ag) + \alpha(g, g)].$$

The condition that $S_\alpha(g_\alpha + \lambda g)$ be a minimum at $\lambda = 0$ yields

$$(Ag_\alpha - f, Ag) + \alpha(g_\alpha, g) = 0 \quad \forall g.$$

This is equivalent to

$$(A^*(Ag_\alpha - f), g) + a(g_\alpha, g) = 0,$$

which implies

$$(A^*A + \alpha I)g_\alpha = A^*f. \quad \blacksquare \tag{2.24}$$

Notice that in the preceding proof we have introduced the operators A^* (adjoint of A) and I (identity). The operator A^* is defined as the unique operator satisfying the bilinear form $(Au, v) = (u, A^*v)$ for every pair of smooth functions u and v in $L^2(I)$.

For example, if $Ag = f$ represents the Volterra integral equation of the first kind

$$f(t) = \int_0^t k(t - s)g(s)\,ds, \qquad 0 \le t \le T,$$

it is a simple task to verify that the adjoint operator is given by

$$A^*v(t) = \int_t^T k(s - t)v(s)\,ds, \qquad 0 \le t \le T.$$

There might be situations where one would like to emphasize the physical meaning of the adjoint problem. In this sense, we point out that the system of canonical equations (2.24) is equivalent to the weakly coupled system of equations

$$\begin{aligned} Ag_\alpha - \sqrt{\alpha}\,v_\alpha &= f, \\ A^*v_\alpha + \sqrt{\alpha}\,g_\alpha &= 0, \end{aligned} \qquad (2.25)$$

where v_α is the "adjoint" variable.

Elimination of v_α from system (2.25) gives (2.24) and, starting from

$$(A^*A + \alpha I)g_\alpha = A^*f,$$

we have

$$A^*(Ag_\alpha - f) + \alpha g_\alpha = 0. \qquad (2.26)$$

Defining

$$\sqrt{\alpha}\,v_\alpha = Ag_\alpha - f, \qquad (2.27)$$

we obtain, using equality (2.26),

$$A^*\left(\sqrt{\alpha}\,v_\alpha\right) + \alpha g_\alpha = 0,$$

which is equivalent to

$$A^*v_\alpha + \sqrt{\alpha}\,g_\alpha = 0. \qquad (2.28)$$

Equations (2.27) and (2.28) define system (2.24). ∎

It is interesting to note that if $\alpha = 0$, system (2.24) collapses into the uncoupled system of linear equations

$$\begin{aligned} Ag &= f, \\ A^*v &= 0. \end{aligned} \qquad (2.29)$$

If the only solution of the homogeneous adjoint equation $A^*v = 0$ is $v \equiv 0$, then solving (2.29) is equivalent to solving the original single problem $Ag = f$. This statement—Fredholm alternative—can be seen from the following argument.

From $Ag = f$, we get

$$(v, Ag) = (v, f)$$

and from $A^*v = 0$, we get

$$(g, A^*v) = (g, 0) = 0. \tag{2.30}$$

On the other hand,

$$(v, Ag) = (A^*v, g) = (g, A^*v) \tag{2.31}$$

using the definition of A^* and the commutativity of the inner product.

Hence, from (2.30) and (2.31), we have

$$(v, f) = 0.$$

The conclusion is that any solution of the homogeneous equation $A^*v = 0$ has to be orthogonal to the data function of the equation $Ag = f$. However, if the only solution of $A^*v = 0$ is $v \equiv 0$, there is no such restriction.

Another important aspect of system (2.25) is that it strongly suggests being solved by the method of successive approximations. System (2.25) is equivalent to

$$Ag_\alpha - \sqrt{\alpha}\, v_\alpha = f,$$
$$\sqrt{\alpha}\, \beta A^* v_\alpha + \alpha\beta g_\alpha - g_\alpha + g_\alpha = 0,$$

where β is an arbitrary nonzero real number to be determined. We elect β to depend on the iteration and rewrite the preceding system as

$$Ag_\alpha^{(n)} - \sqrt{\alpha}\, v_\alpha^{(n)} = f,$$
$$\sqrt{\alpha}\, \beta_n A^* v_\alpha^{(n)} + \alpha\beta_n g_\alpha^{(n)} - g_\alpha^{(n)} + g_\alpha^{(n+1)} = 0,$$

to obtain

$$\sqrt{\alpha}\, v_\alpha^{(n)} = Ag_\alpha^{(n)} - f,$$
$$g_\alpha^{(n+1)} = g_\alpha^{(n)} - \beta_n\left[\alpha g_\alpha^{(n)} + A^* \sqrt{\alpha}\, v_\alpha^{(n)}\right], \tag{2.32}$$

$$n = 0, 1, \ldots, \quad g_\alpha^{(0)} \text{ arbitrary, usually 0.}$$

Remarks

1. Each iteration in (2.32) involves the solution of two "direct" problems: one corresponding to the original operator, $Ag_\alpha^{(n)}$, and the other associated with the adjoint operator, $A^*(\sqrt{\alpha}\, v_\alpha^{(n)})$.
2. The gradient of the functional (2.23) is given by

$$\nabla S_\alpha(g) = \alpha g + A^*(Ag - f)$$

and it is easily computed if the solution of the adjoint problem is known. In fact, taking into consideration (2.27), we can write $\nabla S_\alpha(g) = \alpha g + A^*(\sqrt{\alpha}\, v)$, and for each iteration we get

$$\nabla S_\alpha(g_\alpha^{(n)}) = \alpha g_\alpha^{(n)} + A^*(\sqrt{\alpha}\, v_\alpha^{(n)}). \tag{2.33}$$

These considerations allow us to choose β_n, for each n, in such a manner that system (2.32) can now be solved by the conjugate gradient method.

The complete abstract algorithm for the regularization–adjoint–conjugate gradient method—Method IV—is as follows:
For $n = 0$,

Step 0. Set $g_\alpha^{(0)} = 0$ and choose $\alpha > 0$.

Step 1. Compute $Ag_\alpha^{(0)}$, that is, solve the original direct problem.

Step 2. Compute the residual $\sqrt{\alpha}\, v_\alpha^{(0)} = Ag_\alpha^{(0)} - f_m$.

Step 3. Compute $A^*(\sqrt{\alpha}\, v_\alpha^{(0)})$, that is, solve the direct adjoint problem.

Step 4. Evaluate the gradient $d_0 = \nabla S_\alpha(g_\alpha^{(0)})$ using formula (2.33).

Step 5. Set $r_0 = \dfrac{\|d_0\|}{\alpha\|d_0\|^2 + \|Ad_0\|^2}$.

Step 6. Update $g_\alpha^{(1)} = g_\alpha^{(0)} - r_0 d_0$.

For $n = 1, 2, \ldots$,

Step 1′. Solve the original direct problem $Ag_\alpha^{(n)}$.

Step 2′. Compute the residual $\sqrt{\alpha}\, v_\alpha^{(n)} = Ag_\alpha^{(n)} - f_m$.

Step 3′. Solve the direct adjoint problem $A^*(\sqrt{\alpha}\, v_\alpha^{(n)})$.

Step 4′. Evaluate the gradient $d_0 = \nabla S_\alpha(g_\alpha^{(n)})$ using formula (2.33).

Step 4″. Compute $d_n = \nabla S_\alpha(g_\alpha^{(n)}) + \dfrac{\left\|\nabla S_\alpha(g_\alpha^{(n)})\right\|^2}{\left\|\nabla S_\alpha(g_\alpha^{(n-1)})\right\|^2}$.

Step 5′. Set $r_n = \dfrac{\left(\nabla S_\alpha(g_\alpha^{(n-1)}), d_n\right)}{\alpha\|d_n\|^2 + \|Ad_n\|^2}$.

Step 6′. Update $g_\alpha^{(n+1)} = g_\alpha^{(n)} - r_n d_n$.

Alternatively, the noniterative algorithm for Tikhonov's method can be described by the abstract algorithm:

Step 1. Choose $\alpha > 0$.
Step 2. Solve the canonical system of linear equations $(A^*A + \alpha I)g_\alpha^\varepsilon = A^*f_m$.

As is the case with Methods I, II, and III, it follows from well-known estimates in the theory of Tikhonov's regularization—see the references at the end of this chapter—that

$$\|g - g_\alpha\| = C_0\sqrt{\alpha} \qquad \text{(Consistency)}, \qquad (2.34)$$

for some constant $C_0 > 0$, independent of α; and

$$\|g_\alpha - g_\alpha^\varepsilon\| \le \frac{\varepsilon}{\sqrt{\alpha}} \qquad \text{(Stability)}. \qquad (2.35)$$

Choosing $\alpha = O(\varepsilon)$ and combining (2.34) and (2.35), we have

$$\|g - g_\alpha^\varepsilon\| = O(\sqrt{\varepsilon}),$$

which shows that, theoretically, as the quality of the data improves ($\varepsilon \to 0$), we get convergence with rate $O(\sqrt{\varepsilon})$.

We are now ready for the finite-dimensional versions of Methods I to IV.

2.3   NUMERICAL IMPLEMENTATIONS

Method I

To numerically approximate $g_\delta^\varepsilon(t)$, a quadrature formula for the convolution equation (2.11) is needed. In order to simplify the discrete error analysis, we shall further assume the data function $f_m(t)$ to be uniformly Lipschitz on $I = [0, 1]$, with Lipschitz constant L.

Since, in practice, only a discrete set of data points is available, for $h > 0$ and $Nh = 1$, we let $t_i = ih$ and denote $f_m(t_i) = f_{m,i}$, $i = 0, 1, \ldots, N$, with $f_{m,0} = 0$. The objective is to introduce a simple quadrature formula and avoid any artificial smoothing in the process. To that effect, given t_j, $j = 0, 1, \ldots, N$, we define

$$p_m(t) = \sum_{i=0}^{j} f_{m,i}\varphi_i(t), \qquad 0 \le t \le t_j,$$

a piecewise interpolant of $f_m(t)$ at the grid points t_j. Here,

$$\varphi_0(t) = \begin{cases} 1, & \text{if } 0 \le t \le h/2, \\ 0, & \text{otherwise,} \end{cases} \qquad \varphi_j(t) = \begin{cases} 1, & \text{if } t_j - h/2 \le t \le t_j, \\ 0, & \text{otherwise,} \end{cases}$$

and

$$\varphi_i(t) = \begin{cases} 1, & \text{if } t_i - h/2 \le t \le t_i + h/2, \\ 0, & \text{otherwise,} \end{cases} \qquad i = 1, 2, \ldots, j - 1.$$

Once the data function has been extended and the radius of mollification selected—as indicated in Sec. 2.2—we evaluate

$$
\begin{aligned}
(\rho_\delta * p_m)(t_i) &= \sum_k f_{m,k}(\rho_\delta * \varphi_k)(t_i) \\
&= \sum_k f_{m,k}\sigma_{\delta,k,i} \\
&= \beta_{m,\delta,i}, \qquad i = 0, 1, \ldots, N, \qquad (2.36)
\end{aligned}
$$

with the weights $\sigma_{\delta,k,i}$ computed exactly.

Recalling the Lipschitz property of f_m and the fact that p_m is a piecewise constant interpolant of f_m at the grid points, we have

$$\|f_m - p_m\|_{\infty, I} = \max_{0 \le t \le 1} |f_m(t) - p_m(t)| \le Lh$$

or

$$f_m(t) = p_m(t) + O(h).$$

Hence,

$$(\rho_\delta * f_m)(t) = (\rho_\delta * p_m)(t) + O(h), \qquad 0 \le t \le 1, \qquad (2.37)$$

and the smoothness of $(\rho_\delta * p_m)(t)$, repeating the previous argument, implies that

$$\sum_{i=0}^{N} \beta_{m,\delta,i}\varphi_i(t) = (\rho_\delta * p_m)(t) + O(h), \qquad 0 \le t \le 1.$$

On the other hand, the smoothness of $(\rho_\delta * f_m)'$ ensures

$$(\rho_\delta * f_m)'(t) = \sum_{i=0}^{N} (\rho_\delta * f_m)'(t_i)\varphi_i(t) + O(h), \qquad 0 \le t \le 1, \quad (2.38)$$

and also

$$(\rho_\delta * f_m)'(t_0) = \frac{1}{h}\left[(\rho_\delta * f_m)(t_0) - (\rho_\delta * f_m)(t_1)\right] + O(h),$$

$$(\rho_\delta * f_m)'(t_i) = \frac{1}{2h}\left[(\rho_\delta * f_m)(t_{i+1}) - (\rho_\delta * f_m)(t_{i-1})\right] + O(h^2),$$

$$i = 1, 2, \ldots, j - 1, \quad (2.39)$$

and

$$(\rho_\delta * f_m)'(t_j) = \frac{1}{h}\left[(\rho_\delta * f_m)(t_j) - (\rho_\delta * f_m)(t_{j-1})\right] + O(h),$$

$$j = 1, 2, \ldots, N.$$

From (2.38), we have

$$(\rho_\delta * f_m)'(t) = \sum_{i=0}^{j} (\rho_\delta * f_m)'(t_i)\varphi_i(t) + O(h),$$

$$0 \le t \le t_j, j = 1, 2, \ldots, N.$$

Using expressions (2.37) and (2.39) in the last equality,

$$(\rho_\delta * f_m)'(t) = \frac{1}{h}\left[(\rho_\delta * f_m)(t_0) - (\rho_\delta * f_m)(t_1)\right]\varphi_0(t)$$

$$+ \frac{1}{2h}\sum_{i=0}^{j-1}\left[(\rho_\delta * p_m)(t_{i+1}) - (\rho_\delta * p_m)(t_{i-1})\right]\varphi_i(t)$$

$$+ \frac{1}{h}\left[(\rho_\delta * f_m)(t_j) - (\rho_\delta * f_m)(t_{j-1})\right]\varphi_j(t) + O(h),$$

$$0 \le t \le t_j, j = 0, 1, \ldots, N.$$

Inserting this expression into (2.11) and integrating exactly the factors

$$w_i^j = \frac{1}{\sqrt{h}}\int_0^{t_j}\varphi_i(t)(t_j - t)^{-1/2}\, dt, \quad i, j = 0, 1, \ldots, N, \quad (2.40)$$

we get

$$g_\delta^\varepsilon(t_j) = g_{\delta,h}^\varepsilon(t_j) + O(\sqrt{\varepsilon}), \quad j = 0, 1, \ldots, N, \quad (2.41)$$

where

$$g_{\delta,h}^{\varepsilon}(0) = 0,$$

$$g_{\delta,h}^{\varepsilon}(t_j) = \frac{1}{\pi\sqrt{h}} \left\{ [\beta_{m,\delta,0} - \beta_{m,\delta,1}]w_0^j + \sum_{i=1}^{j-1} [\beta_{m,\delta,i+1} - \beta_{m,\delta,i-1}]w_1^j \right.$$

$$\left. + [\beta_{m,\delta,j} - \beta_{m,\delta,j-1}]w_j^j \right\}, \qquad j = 0, 1, \ldots, N, \quad (2.42)$$

taking into account (2.36).

The approximate inverse Abel transform is computed with formula (2.42) after evaluating the factors defined in (2.36) and (2.40).

Combining estimate (2.41) with the restriction to the grid points of the error estimate (2.12) and using the triangle inequality, we have the following theorem.

Theorem 2.4 (Discrete Error Estimate)   Under the conditions of Theorem 2.2, if $f_m(t)$ is uniformly Lipschitz on $I = [0, 1]$ with Lipschitz constant L and if the approximate solution $g_{\delta,h}^{\varepsilon}(t_j)$ is computed using (2.42), then

$$\|g - g_{\delta,h}^{\varepsilon}\|_{\infty,I} \leq \frac{2}{\pi} \left[3\delta M_2 + \frac{2}{\sqrt{\pi}} \frac{\varepsilon}{\delta} \right] + C(L, \delta, \varepsilon)\sqrt{h} ,$$

where

$$C(L, \delta, \varepsilon) = C\left(L, \left\| \frac{d}{dx}(\rho_\delta * p_m) \right\|_{\infty,I}, \left\| \frac{d^{(i)}}{dx^{(i)}}(\rho_\delta * f_m) \right\|_{\infty,I} \right), \qquad i = 1, 2, 3.$$

The error estimate for the discrete case is obtained by adding the global truncation error to the error estimate of the nondiscrete case.

Method II

The convolution $\rho_\mu * k$ requires an extension of the singular kernel k for values of t less than or equal to 0. In our numerical implementation we use the following symmetric extension:

$$k(0) = \frac{2}{\sqrt{h}}, \qquad k(t) = k(-t), t < 0.$$

The discrete approximation is now straightforward.

Writing $s_j = (\rho_\mu * k)(t_j)$, $j = 0, 1, \ldots, N$, the discrete convolution formulas corresponding to (2.13) are as follows with $g^\varepsilon_{\mu, h}(t_j)$ indicating the approximate inverse Abel transform at the grid points.

$$g^\varepsilon_{\mu, h}(0) = 0,$$

$$g^\varepsilon_{\mu, h}(t_j) = \frac{1}{2\pi} \sum_{k=1}^{j} s_{j-k}(f_{m, k+1} - f_{m, k-1}), \qquad j = 1, 2, \ldots, N - 1,$$

and

$$g^\varepsilon_{\mu, h}(1) = g^\varepsilon_{\mu, h}(t_{N-1}) + \frac{1}{\pi} s_0(f_{m, N} - f_{m, N-1}).$$

Method III

In this case, given t_j, $j = 0, 1, \ldots, N$, we define

$$p_m(t) = \sum_{i=0}^{j} f_{m, i} \phi_i(t), \qquad 0 \le t \le t_j,$$

a piecewise linear interpolant of $f_m(t)$ at the grid points t_j. Here,

$$\phi_0(t) = \begin{cases} 1 - t/h, & 0 \le t \le h, \\ 0, & \text{otherwise}, \end{cases}$$

$$\phi_j(t) = \begin{cases} 1 + (t - t_j)/h, & t_{j-1} \le t \le t_j, \\ 0, & \text{otherwise}, \end{cases}$$

and

$$\phi_i(t) = \begin{cases} 1 + (t - t_i)/h, & t_{i-1} \le t \le t_i, \\ 1 - (t - t_i)/h, & t_i \le t \le t_{i+1}, \\ 0, & \text{otherwise}, \end{cases}$$

$$i = 1, 2, \ldots, j - 1, 1 \le j \le N.$$

The quadrature formula is obtained by directly convolving the kernel function H_γ against p_m as follows:

$$g^\varepsilon_{\gamma, h}(t_j) = \frac{1}{\pi}(H_\gamma * p_m)(t_j)$$

$$= \frac{1}{\pi} \int_0^{t_j} H_\gamma(t_j - s) p_m(s) \, ds$$

$$= \frac{1}{\pi} \sum_{i=0}^{j} f_{m, i} b_{\gamma, i}(t_j),$$

where

$$b_{\gamma,i}(t_j) = \int_0^{t_j} H_\gamma(t_j - s)\phi_i(s)\, ds$$

is evaluated exactly for $i = 0, 1, \ldots, j$.

Thus, we have

$$g_{\gamma,h}^\varepsilon(0) = 0,$$

$$g_{\gamma,h}^\varepsilon(t_j) = \frac{1}{\pi} \sum_{i=0}^{j} f_{m,i} b_{\gamma,i}(t_j), \qquad j = 1, 2, \ldots, N. \tag{2.43}$$

From (2.18) and (2.43), it follows that

$$\left| g_{\gamma,h}^\varepsilon(t_j) - g_\gamma^\varepsilon(t_j) \right| = \frac{1}{\pi} \left| \int_0^t H_\gamma(t_j - s)(f_m(s) - p_m(s))\, ds \right|$$

$$\leq \frac{1}{\pi} \| f_m - p_m \|_{\infty, I} \int_0^{t_j} \left| H_\gamma(t_j - s) \right| ds. \tag{2.44}$$

From the definition (2.18) of the kernel function H_γ, we immediately obtain

$$\int_0^{t_j} \left| H_\gamma(t_j - s) \right| ds = 2\gamma^{-1/2} - t^{-1/2} \leq 2\gamma^{-1/2}. \tag{2.45}$$

Recalling the Lipschitz property of f_m and the fact that p_m is a piecewise interpolant of f_m at the grid points, we have

$$\max_{0 \leq t \leq t_j} \left| f_m(t) - p_m(t) \right| = \max_{i=1,2,\ldots,N} \left\{ \max_{t_{i-1} \leq t \leq t_i} \left| f_m(t) - p_m(t) \right| \right\} \leq Lh. \tag{2.46}$$

Using the estimates (2.45) and (2.46) in inequality (2.44), we get

$$\left| g_\gamma^\varepsilon(t_j) - g_{\gamma,h}^\varepsilon(t_j) \right| \leq \frac{2}{\pi} L \frac{h}{\sqrt{\gamma}}, \qquad j = 0, 1, \ldots, N.$$

Combining this upper bound with the restriction to the grid points of the error estimate of Theorem 2.3, we have the following proposition.

Theorem 2.5 (Discrete Error Estimate)   Under the conditions of Theorem 2.3, if $f_m(t)$ is uniformly Lipschitz on $I = [0, 1]$ with Lipschitz constant L and if the approximate solution of Abel's inverse problem $g_{\gamma,h}^\varepsilon(t_j)$ is computed using (2.43), then

$$\max_{0 \le j \le N} \left| g_\gamma^\varepsilon(t_j) - g_{\gamma,h}^\varepsilon(t_j) \right| \le \frac{5}{2\pi} M_1 \sqrt{\gamma} + \frac{2}{\pi} \frac{1}{\sqrt{\gamma}} (\varepsilon + Lh) + O(\gamma^{3/2}).$$

The error estimate for the discrete case is obtained by adding the global truncation error to the error estimate of the nondiscrete case.

Method IV

Iterative Algorithm

Discretization leads to a finite-dimensional version of the combined regularization–adjoint–conjugate gradient algorithm of Sec. 2.2. The operators A and A^* are now represented by an $N \times N$ matrix A and its transpose A^T, respectively. The approximate discrete solution $g_{\alpha,m}^{(n)}$, obtained after n iterations, the gradient $\nabla S_\alpha(g_{\alpha,m}^{(n)})$, d_n, r_n, $g_{\alpha,m}^{(0)}$, and the residual $\alpha v_{\alpha,m}^{(n)}$ are now N-dimensional real vectors. From (2.2), a simple discretization gives the lower triangular system of linear equations

$$h \sum_{i=1}^{j} a_{j+1-i} g_{\alpha,m,i}^{(n)} = f_{m,i},$$

where

$$a_j = \frac{1}{\sqrt{jh}}, \qquad j = 1, 2, \ldots, N,$$

indicates the $(j - 1)$ subdiagonal of the $N \times N$ matrix A.

The discrete algorithm for the conjugate gradient method follows exactly the steps described previously in Sec. 2.2, and we only have to add the necessary stopping criteria, given by

$$\left\| g_{\alpha,m}^{(n)} - g_{\alpha,m}^{(n-1)} \right\|_{2,I} \le \text{TOL} \left\| g_{\alpha,m}^{(n)} \right\|_{2,I},$$

where TOL is a small positive tolerance parameter entered by the user and

$$\|f\|_{2,I} = \left\{ \frac{1}{N+1} \sum_{j=0}^{N} [f_j]^2 \right\}^{1/2} \tag{2.47}$$

is the discrete—weighted—l^2 norm on $I = [0, 1]$.

Direct Algorithm

The finite-dimensional version of the canonical equations (2.24) gives the system of N linear equations with N unknowns

$$(A^T A + \alpha I) \mathbf{y}_\alpha = A^T \mathbf{b}, \tag{2.48}$$

with the $N \times N$ matrix A and its transpose A^T as described previously, I the $N \times N$ identity matrix, $\mathbf{y}_\alpha = (y_1^\alpha, y_2^\alpha, \ldots, y_N^\alpha)^T$, and $\mathbf{b} = (f_1, f_2, \ldots, f_N)^T$. We assume that the vector $\mathbf{y}$ is the unique solution of the least squares problem

$$A^T A \mathbf{y} = A^T \mathbf{b}. \tag{2.49}$$

In other words, we assume that the symmetric matrix $A^T A$ is positive definite and, consequently, invertible. Moreover, the eigenvalues $\{\lambda_i\}_{i=1}^N$ of $A^T A$ satisfy

$$0 < \lambda_1 \le \lambda_2 \le \cdots \le \lambda_N.$$

However, the matrices $A^T A$ associated with discretized models of ill-posed problems are, most of the time, nearly numerically singular; the smallest eigenvalue, λ_1, is "too close to 0." Equivalently, we say that the system (2.49) is highly ill conditioned for inversion or that small perturbations in the data vector $\mathbf{b}$ might produce gigantic changes in the solution vector $\mathbf{y}$, or that the condition number of the matrix $A^T A$ is very large.

The condition number of a matrix depends on the norm being used and for symmetric matrices it is common to consider the l^2-norm condition number given by

$$k_2(A^T A) = \frac{\lambda_N}{\lambda_1}.$$

We would like to investigate how sensitive the solution of system (2.48) is to changes in the data vector $\mathbf{b}$. For instance, if $\|\mathbf{b} - \mathbf{b}_m\|_{\infty, I} \le \varepsilon$, what can we say about $\|\mathbf{y}_\alpha - \mathbf{y}_{\alpha, m}\|_{\infty, I}$? [Notice that we always have $\|\mathbf{y}_\alpha - \mathbf{y}_{\alpha, m}\|_{2, I} \le \|\mathbf{y}_\alpha - \mathbf{y}_{\alpha, m}\|_{\infty, I}$ when using the weighted l^2 norm (2.47).]

Let us begin by looking at the partitioned $2N \times 2N$ symmetric eigenvalue problem

$$\begin{pmatrix} 0 & A \\ A^T & 0 \end{pmatrix} \begin{pmatrix} \mathbf{u} \\ \mathbf{v} \end{pmatrix} = \lambda \begin{pmatrix} \mathbf{u} \\ \mathbf{v} \end{pmatrix}.$$

It follows immediately that

$$A\mathbf{v} = \lambda \mathbf{u} \tag{2.50}$$

and

$$A^T \mathbf{u} = \lambda \mathbf{v}. \tag{2.51}$$

Hence,

$$A^T A \mathbf{v} = \lambda^2 \mathbf{v} \quad \text{and} \quad A A^T \mathbf{u} = \lambda^2 \mathbf{u} \tag{2.52}$$

Consider the set of orthonormal eigenvectors $\{\mathbf{u}_{(i)}\}_{i=1}^{N}$ and $\{\mathbf{v}_{(i)}\}_{i=1}^{N}$ associated with the symmetric matrices AA^T and A^TA, respectively.

The vector $\mathbf{y}_\alpha$ solution of (2.48) can be uniquely represented as a linear combination of the eigenvectors $\mathbf{v}_{(i)}$. For example, we can write

$$\mathbf{y}_\alpha = \sum_{i=1}^{N} y_i^\alpha \mathbf{v}_{(i)}, \qquad y_i^\alpha = (\mathbf{y}_\alpha)^T \mathbf{v}_{(i)}, \qquad i = 1, 2, \ldots, N.$$

Thus,

$$A^TA\mathbf{y}_\alpha = \sum_{i=1}^{N} y_i^\alpha A^TA\mathbf{v}_{(i)} = \sum_{i=1}^{N} y_i^\alpha \lambda_i^2 \mathbf{v}_{(i)},$$

$$\alpha I \mathbf{y}_\alpha = \sum_{i=1}^{N} \alpha y_i^\alpha \mathbf{v}_{(i)},$$

and

$$(A^TA + \alpha I)\mathbf{y}_\alpha = \sum_{i=1}^{N} (\lambda_i^2 + \alpha) y_i^\alpha \mathbf{v}_{(i)}, \tag{2.53}$$

using (2.52).

The data vector $\mathbf{b}$ can also be uniquely represented as a linear combination—this time—of the eigenvectors $\mathbf{u}_{(i)}$, that is,

$$\mathbf{b} = \sum_{i=1}^{N} b_i \mathbf{u}_{(i)}, \qquad b_i = \mathbf{b}^T \mathbf{u}_{(i)}, \qquad i = 1, 2, \ldots, N.$$

Then

$$A^T\mathbf{b} = \sum_{i=1}^{N} b_i A^T\mathbf{u}_{(i)} = \sum_{i=1}^{N} b_i \lambda_i \mathbf{v}_{(i)} \tag{2.54}$$

by equality (2.51). From (2.53) and (2.54), using (2.48), it follows that

$$(\lambda_i^2 + \alpha) y_i^\alpha = \lambda_i b_i, \qquad i = 1, 2, \ldots, N,$$

or

$$y_i^\alpha = \frac{\lambda_i b_i}{\alpha + \lambda_i^2} = \frac{\lambda_i}{\alpha + \lambda_i^2} \mathbf{b}^T \mathbf{u}_{(i)}, \qquad i = 1, 2, \ldots, N. \tag{2.55}$$

Notice that if in (2.55) we consider $\alpha = 0$, we get a characterization for the solution vector **y** of the system (2.49). Indeed, we obtain

$$y_i = \frac{1}{\lambda_i} \mathbf{b}^T \mathbf{u}_{(i)}, \qquad i = 1, 2, \ldots, N. \tag{2.56}$$

The effect of the "regularization" can be seen very clearly now by comparing (2.55) and (2.56).

If the data vector **b** in (2.48) is replaced by the noisy data vector $\mathbf{b}_m$, we obtain, in the same manner,

$$(y_m^\alpha)_i = \frac{\lambda_i}{\alpha + \lambda_i^2} (\mathbf{b}_m)^T \mathbf{u}_{(i)}, \qquad i = 1, 2, \ldots, N. \tag{2.57}$$

Once again, we are facing a familiar situation.

From (2.55) and (2.56), we have

$$|y_i - y_i^\alpha| = \left| \frac{1}{\lambda_i} - \frac{\lambda_i}{\alpha + \lambda_i^2} \right| \left| \mathbf{b}^T \mathbf{u}_{(i)} \right| = \frac{\alpha}{\lambda_i(\alpha + \lambda_i^2)} \left| \mathbf{b}^T \mathbf{u}_{(i)} \right|$$

$$\leq \frac{\alpha}{\lambda_1^3} \|\mathbf{b}\|_{2, I} \|\mathbf{u}_{(i)}\|_{2, I}$$

$$= \frac{\alpha}{\lambda_1^3} \|\mathbf{b}\|_{2, I},$$

which implies

$$\|\mathbf{y} - \mathbf{y}_\alpha\|_{\infty, I} \leq \frac{\alpha}{\lambda_1^3} \|\mathbf{b}\|_{\infty, I}, \tag{2.58}$$

and we have *consistency*. In the absence of noise in the data vector, as $\alpha \to 0$, $\|\mathbf{y} - \mathbf{y}_\alpha\|_{\infty, I} \to 0$. ■

For stability, we compare (2.55) and (2.57). We get

$$\left| y_i^\alpha - (y_m^\alpha)_i \right| = \frac{\lambda_i}{\alpha + \lambda_i^2} \left| \mathbf{b}^T \mathbf{u}_{(i)} - (\mathbf{b}_m)^T \mathbf{u}_{(i)} \right|$$

$$\leq \frac{\lambda_N}{\alpha} \|\mathbf{b} - \mathbf{b}_m\|_{\infty, I}$$

$$\leq \frac{\lambda_N}{\alpha} \varepsilon, \qquad i = 1, 2, \ldots, N,$$

or

$$\|\mathbf{y}_\alpha - \mathbf{y}_{\alpha, m}\|_{\infty, I} \le \frac{\lambda_N}{\alpha}\varepsilon. \quad \blacksquare \qquad (2.59)$$

Thus, the method is *stable*. For fixed $\alpha > 0$, as $\varepsilon \to 0$, $\|\mathbf{y}_\alpha - \mathbf{y}_{\alpha, m}\|_{\infty, I} \to 0$. As usual, from inequalities (2.58) and (2.59), we obtain

$$\|\mathbf{y} - \mathbf{y}_{\alpha, m}\|_{\infty, I} \le \frac{\alpha}{\lambda_1^3}\|\mathbf{b}\|_{\infty, I} + \frac{\lambda_N}{\alpha}\varepsilon, \qquad (2.60)$$

and if $\alpha = O(\sqrt{\varepsilon})$, as $\varepsilon \to 0$, we have convergence—in the uniform norm—with rate $O(\sqrt{\varepsilon})$. $\quad \blacksquare$

Remarks

1. If a lower uniform bound—independent of N—is available for λ_1 and if λ_N is known as a function of the dimension N, it is possible to estimate the error between the assumed unique solution function g of the operator equation $A^*Ag = A^*f$ and the N-dimensional solution vector $\mathbf{y}_{\alpha, m}$ given by (2.48) with the vector $\mathbf{b}$ replaced by the vector $\mathbf{b}_m$. Setting $\alpha = O(1/N)$ in (2.60) and selecting N as a function of ε—minimizing the upper bound for the error term—we can obtain convergence as $\varepsilon \to 0$. [The uniform lower bound for $\lambda_1 = \lambda_1(N)$ is generally provided by the knowledge—when available—of the first eigenvalue of the operator A^*A.]

2. The analysis in the finite-dimensional setting mimics exactly the situation in the infinite-dimensional case. This is always the case when we discretize the continuous—nondiscrete—model. However, we should keep in mind that there are situations where the physical model is inherently discrete in nature and better adapted—directly—to computer analysis than its continuous approximation.

3. The symmetric positive definite matrix $Q_\alpha = A^TA + \alpha I$, $\alpha > 0$, induces a family of new norms in $\mathbb{R}^N$ as follows:

$$\|\mathbf{v}\|_{Q_\alpha}^2 = (Q_\alpha \mathbf{v})^T \mathbf{v}.$$

It is relative to this norm that the error estimates for the iterations in the conjugate gradient method are presented. It is possible to show—see the references at the end of this chapter—that

$$\|g_{\alpha, m}^{(n)} - g_{\alpha, m}\|_{Q_\alpha, I} \le 2\|g_{\alpha, m}^{(0)} - g_{\alpha, m}\|_{Q_\alpha, I}\left[\frac{[k_2(Q_\alpha)]^{1/2} - 1}{[k_2(Q_\alpha)]^{1/2} + 1}\right]^n.$$

2.4 NUMERICAL RESULTS AND COMPARISONS

In this section we describe some tests that we have implemented in order to compare the performance of the methods introduced in the previous sections.

We tested the methods on three examples. In all of them, the exact data function is denoted $f(t)$ and the noisy data function $f_m(t)$ is obtained by adding an ε random error to $f(t)$, that is, $f_m(t_j) = f(t_j) + \varepsilon\sigma_j$, where $t_j = jh$, $j = 0, 1, \ldots, N$, $Nh = 1$, and σ_j is a uniform random variable with values in $[-1, 1]$ such that

$$\max_{0 \leq j \leq N} \left| f_m(t_j) - f(t_j) \right| \leq \varepsilon.$$

The exact inverse Abel transform is denoted $g(t)$ and its approximation—given by any of the methods—is denoted $g^{\varepsilon}_{p, h}(t_j)$, where p repre-

TABLE 2.1. Error Norms as Functions of ε
(Example 1 with $N = 500$)

Method	Parameter	$\varepsilon = 0.000$	$\varepsilon = 0.005$	$\varepsilon = 0.010$
I	$\delta = 0.0080$	0.0000	0.0048	0.0096
II	$\mu = 0.0080$	0.0005	0.0137	0.0274
III	$\gamma = 0.0040$	0.0302	0.0315	0.0349
IV	$\alpha = 0.0064$	0.0279	0.0294	0.0359

TABLE 2.2. Error Norms as Functions of ε
(Example 2 with $N = 500$)

Method	Parameter	$\varepsilon = 0.000$	$\varepsilon = 0.005$	$\varepsilon = 0.010$
I	$\delta = 0.0080$	0.0001	0.0048	0.0096
II	$\mu = 0.0080$	0.0005	0.0136	0.0273
III	$\gamma = 0.0010$	0.0174	0.0263	0.0431
IV	$\alpha = 0.0064$	0.0275	0.0293	0.0365

TABLE 2.3. Error Norms as Functions of ε
(Example 3 with $N = 500$)

Method	Parameter	$\varepsilon = 0.000$	$\varepsilon = 0.005$	$\varepsilon = 0.010$
I	$\delta = 0.0080$	0.0052	0.0052	0.0053
II	$\mu = 0.0080$	0.0295	0.0330	0.0411
III	$\gamma = 0.0010$	0.0648	0.0678	0.0760
IV	$\alpha = 0.0064$	0.0615	0.0618	0.0641

sents the regularization parameter of the particular method; that is, $p = \delta$ for Method I, $p = \mu$ for Method II, $p = \gamma$ for Method III, and $p = \alpha$ for Method IV.

EXAMPLE 1   As a first example we consider the data function $f(t) = t$ with exact inverse Abel transform $g(t) = 2/\pi\sqrt{t}$. This data function satisfies all the necessary hypotheses for the convergence estimates of Secs. 2.2 and 2.3.

EXAMPLE 2   The data function

$$f(t) = \begin{cases} 2t^2, & 0 \le t < 1/2, \\ 1 - 2(1 - t)^2, & 1/2 \le t \le 1, \end{cases}$$

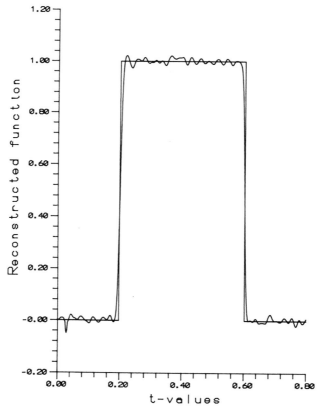

Fig. 2.2 Reconstructed inverse Abel transform. Example 3, Method I: $\varepsilon = 0.01$, $\delta = 0.0080$, $N = 500$.

is only once continuously differentiable on $I = [0, 1]$, partially violating the required conditions for the theoretical analysis of Secs. 2.2 and 2.3. In this example, the exact inverse Abel transform is given by

$$g(t) = \begin{cases} (16/3\pi)t^{3/2}, & 0 \le t < 1/2, \\ (16/3\pi)\left[t^{3/2} + (t - 1/2)^{3/2}\right] - (8/\pi)(t - 1/2)^{1/2}(2t - 1), \\ & 1/2 < t \le 1. \end{cases}$$

EXAMPLE 3 The data function is defined as follows

$$f(t) = \begin{cases} 0, & 0 \le t < 0.2, \\ 2(t - 0.2)^{1/2}, & 0.2 \le t \le 0.6, \\ 2(t - 0.2)^{1/2} - 2(t - 0.6)^{1/2}, & 0.6 < t \le 1. \end{cases}$$

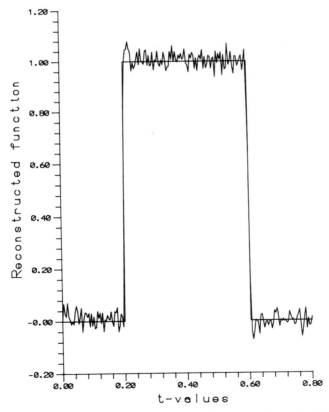

Fig. 2.3 Reconstructed inverse Abel transform. Example 3, Method II: $\varepsilon = 0.01$, $\mu = 0.0080$, $N = 500$.

The first derivative is not continuous on $I = [0, 1]$, strongly violating the necessary hypotheses for the convergence estimates of Secs. 2.2 and 2.3. However, this reconstruction constitutes an important challenging test for the practical utilization of the methods. The exact inverse Abel transform is given by

$$g(t) = \begin{cases} 1, & 0.2 \le t \le 0.6, \\ 0, & \text{otherwise}. \end{cases}$$

The four methods were tested—Method IV in its iterative version—for three different values of N, namely $N = 200$, 500, and 1000, three different values of ε, $\varepsilon = 0.000$, 0.005, and 0.010, and several values of the correspond-

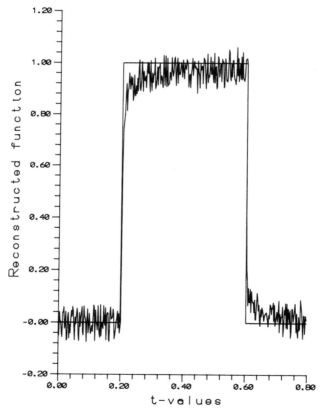

Fig. 2.4 Reconstructed inverse Abel transform. Example 3, Method III: $\varepsilon = 0.01$, $\gamma = 0.0010$, $N = 500$.

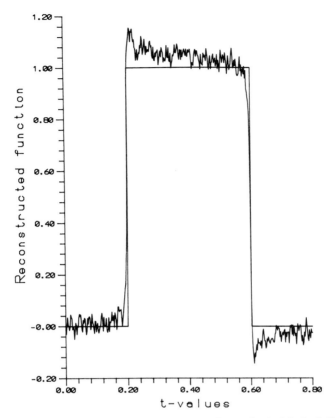

Fig. 2.5 Reconstructed inverse Abel transform. Example 3, Method IV: $\varepsilon = 0.01$, $\alpha = 0.0064$, $N = 500$.

ing regularization parameters that were determined numerically—after several trials—except for Method I where the radius of mollification was selected automatically. These quasi-optimal parameter values are used in the actual tables and figures.

Different values of ε provide a crucial test for stability. Tables 2.1 to 2.3 illustrate this point. The error norms in the tables are computed according to $\|g - g^{\varepsilon}_{p,h}\|_{2,I}$ as defined in (2.47). In the tables, each row corresponds to one of the methods with a fixed regularization parameter value, and shows the change in the error norm due to changes in the level of noise in the data. The numerical results indicate stability. The columns in the tables allow us to compare the performance of the different methods under similar conditions.

Figures 2.2 to 2.5 show the reconstructions of the step function of Example 3 provided by the four methods for the same number of sample data

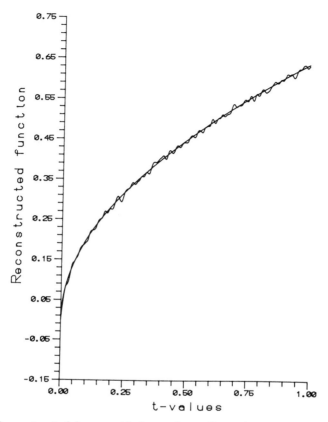

Fig. 2.6 Reconstructed inverse Abel transform. Example 1, Method I: $\varepsilon = 0.001$, $\delta = 0.0080$, $N = 500$.

points, $N = 500$, the same level of noise, $\varepsilon = 0.01$, and quasi-optimal regularization parameters. The qualitative behavior is quite good taking into consideration the high amount of noise in the data. Figures 2.6 and 2.7 show the plots of the computed solutions for Examples 1 and 2, respectively, obtained with Method I for $N = 500$, $\varepsilon = 0.001$, and $\delta = 0.008$.

The consistency and stability of the four methods are clearly confirmed throughout experimentation and very weak dependency on the parameter N is observed. Method I provides an automatic mechanism to select $\delta = \delta(\varepsilon)$ but—as a consequence of the stability of the four methods—it is easy to find, by numerical experimentation, sharp lower and upper bounds for quasi-optimal values of the regularization parameters for Methods II, III, and IV. All the results are very competitive. However, mollification solutions are slightly better in terms of accuracy and Method III—the easiest to implement—seems to be a bit more sensitive to perturbations in the data.

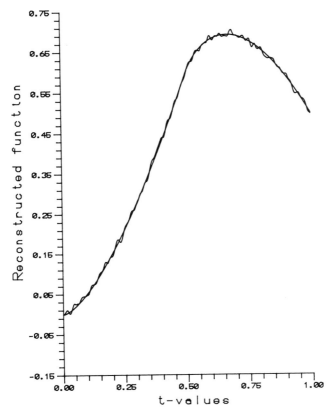

Fig. 2.7 Reconstructed inverse Abel transform. Example 2, Method I: $\varepsilon = 0.005$, $\delta = 0.0080$, $N = 500$.

2.5   EXERCISES

2.1. Show that if $k(t) = 1/\sqrt{t}$, $t > 0$, and $k(t) = k(-t)$, $t < 0$, then $\hat{k}(w) = (1 - i)/2\sqrt{w}$, $i = \sqrt{-1}$.

2.2. Consider the set of canonical equations $(A^*A + \alpha I)g_\alpha = A^*f$, where g_α and f are smooth functions in $L^2(I)$, $I = [0, 1]$, and $\alpha \in \mathbb{R}$, $\alpha > 0$. Show that if $A^*f \neq 0$, then $\|g_\alpha\|_{2, I}$ is a decreasing function of α.

2.3. Under the same conditions of Exercise 2.2, show that $\|Ag_\alpha - f\|_{2, I}$ is an increasing function of α.

2.4. (*Choice of the regularization parameter*, α, *in Tikhonov's method*). If

$$\|Ag_{\bar{\alpha}} - f\|_{2, I} = \varepsilon \qquad \text{(Discrepancy principle)} \qquad (2.61)$$

for $\bar{\alpha} > 0$, show that $g_{\bar{\alpha}}$ is the unique solution of the problem $\text{Min}\|g_\alpha\|^2_{2,I}$ over all possible functions g_α subject to the constraint $\|Ag_\alpha - f\|^2_{2,I} \le \varepsilon^2$, where $g_{\bar{\alpha}}$ is the solution of $(A^*A + \alpha I)g_\alpha = A^*f$ with $\alpha = \bar{\alpha}$. This is a characterization of the solution $g_{\bar{\alpha}}$ obtained using the parameter criterion given by (2.61).

2.5. Given the $N \times N$ system of linear equations $Ax = b$, with $A = A^T$ and $x^TAx > 0$ for $x \ne 0$, the vectors $\{p_{(k)}\}^N_{k=1}$ form a conjugate basis for $\mathbb{R}^N$ if they are a basis for $\mathbb{R}^N$ and $(p_{(i)})^TAp_{(j)} = 0, \forall i \ne j, 1 \le i, j \le N$. Any conjugate direction method to solve $Ax = b$ has the form

$$x^{(k+1)} = x^{(k)} - \beta_k p_{(k)}, \qquad k = 0, 1, \dots, N-1, \qquad x^{(0)} = 0.$$

Show that

$$\beta_k = \frac{(p_{(k)})^T(Ax^{(k)} - b)}{(p_{(k)})^TAp_{(k)}}.$$

2.6. (a) From the fact that an $N \times N$ matrix is singular if and only if its determinant is 0, one might expect that ill-conditioned matrices must have small determinants. Consider $A = A^T$ and find A such that

(1) $\det(A) = 1,$

(2) $k_2(A) = 10^{60}.$

(b) If

$$A = \begin{pmatrix} 0.2161 & 0.1441 \\ 1.2969 & 0.8648 \end{pmatrix} \quad \text{and} \quad b = \begin{pmatrix} 0.1480 \\ 0.8642 \end{pmatrix},$$

an "approximate" solution for $Ax = b$ seems to be

$$\tilde{x} = \begin{pmatrix} 0.9911 \\ -0.4870 \end{pmatrix}.$$

In fact, the residual $\Delta b = A\tilde{x} - b = (0.00000001, -0.00000001)^T$. However, the exact solution is

$$x = \begin{pmatrix} 2 \\ -2 \end{pmatrix}.$$

Explain.

2.7. Solve the canonical equations corresponding to Example 3 of Sec. 2.4 using Choleski's method to take advantage of the symmetry and positive definiteness of the $N \times N$ matrix $A^TA + \alpha I$. Consider $N = 100$, $\alpha = 10^{-1}$, 10^{-4}, and 10^{-6}, and $\varepsilon = 10^{-3}$.

2.8. For $N = 10$, 50, and 100, look at the eigenvalues of the matrix A in Exercise 2.7 (no need to compute anything since A is lower triangular) and compare them with the eigenvalues of A^TA. To find the eigenvalues of A^TA, use your favorite package; if you want to have some fun, program the classical method of Jacobi. Pay particular attention to the variation of the l^2-condition number of A^TA as a function of N.

2.9. Program Methods I, II, III, and IV to solve Example 1 of Sec. 2.4. Try some of the parameter values shown in Table 2.1.

2.6 REFERENCES AND COMMENTS

The following references and comments serve to expand the basic material covered in the corresponding sections.

2.1. The solution of many practical problems is reduced to the inversion of Abel's equation by means of suitable computations. Orthogonal function approaches have been used by

E. L. Kosarev, The numerical solution of Abel's integral equation, *U.S.S.R. Computational Math. and Math. Phys.* **13** (1973), 271–277.

G. N. Minerbo and M. E. Levy, Inversion of Abel's integral equation by means of orthogonal polynomials, *SIAM J. Numer. Anal.* **6** (1969), 598–616.

Product integration techniques were utilized by

K. E. Atkinson, The numerical solution of Abel integral equation by a product trapezoidal method, *SIAM J. Numer. Anal.* **11** (1974), 97–101.

R. F. Cameron and S. McKee, High-accuracy product integration methods for the Abel equation, and their application to a problem in scattering theory, *Internat. J. Numer. Methods in Engrg.* **19** (1983), 1527–1536.

Regularization procedures have been analyzed by

R. S. Anderssen, Stable procedures for the inversion of Abel's equation, *J. Inst. Math. Appl.* **17** (1976), 329–342.

and more recently by

K. M. Hanson, A Bayesian approach to nonlinear inversion: Abel inversion from x-ray attenuation data, *Transport Theory, Invariant Imbedding, and Integral Equations, Lecture Notes in Pure and Applied Mathematics* **115**, P. Nelson et al. (eds.), Marcel Dekker, New York, 1989, pp. 363–378.

For a more general point of view, a systematic treatment of numerical methods for Abel's equations with an extensive and complete bibliography is available in

R. Gorenflo and S. Vesella, *Abel Integral Equations: Analysis and Applications*, *Lecture Notes in Mathematics*, No 1461 Springer-Verlag, Berlin; New York, 1991.

See also chapter 1, sections 1.3–1.6.

P. Linz, *Analytical and Numerical Methods for Volterra Equations*, SIAM, Philadelphia, 1985.

A general review and a comprehensive treatment of Radon's transform and tomography can be found in

F. Natterer, *The Mathematics of Computerized Tomography*, Wiley, Chichester, 1986.

The three-dimensional image reconstruction procedure from cone-beam projections is a field of presently stimulating extensive research. See

B. D. Smith, Image reconstruction from cone-beam projections: necessary and sufficient conditions and reconstruction methods, *IEEE Trans. Med. Imaging* **MI-4** (1985), 14–25.

2.2–2.3. For Method I we follow closely the article

D. A. Murio, Stable numerical inversion of Abel's integral equation by discrete mollification, *Theoretical Aspects of Industrial Design*, *Proceedings in Applied Mathematics* **58**, D. A. Field and V. Komkov (eds.), SIAM, Philadelphia, 1992, pp. 92–104.

Method II was originally introduced by

K. Miller, An optimal method for the x-ray reconstruction problem, preliminary report, *Amer. Math. Soc. Notices* (January 1978), A-161.

K. Miller, New results on reconstruction methods from x-ray projections, Lecture Notes (unpublished), Department of Mathematics, University of Firenze, December 1978, available from the author.

The particular way of dealing with the singularity in the convolution kernel in Method III was first reported by

B. K. P. Horn, Density reconstruction using, arbitrary ray-sampling schemes, *Proc. IEEE* **66** (1978), 551–562.

See also

D. A. Murio, D. Hinestroza, and C. E. Mejía, New stable numerical inversion of Abel's integral equation, *Comput. Math. Appl.* **23** (1992), 3–11.

The generalization to three-dimensional reconstructions is available in a comprehensive review article by

B. D. Smith, Cone-beam tomography: recent advances and a tutorial review, *Opt. Engrg. J.* **29** (1990), 524–534.

The iterative form of Method IV can be found in the paper by

D. A. Murio and C. E. Mejía, Comparison of four stable methods for Abel's integral equation, *Inverse Design Concepts and Optimization*, G. S. Dulikravich (ed.), Washington, DC, 1991, pp. 239–252.

A complete exposition of the conjugate gradient method in the infinite-dimensional setting is presented in

W. M. Patterson, *Iterative Methods for the Solution of Linear Operator Equations in Hilbert Spaces, Lecture Notes in Mathematics* **394**, Springer-Verlag, Berlin; New York, 1974.

Its finite-dimensional counterpart is clearly discussed in

P. G. Ciarlet, *Introduction to Numerical Linear Algebra and Optimisation*, Cambridge University Press, Cambridge, 1989.

For the general theory of the Tikhonov method, see chapter 1, sections 1.3–1.6.

C. W. Groetsch, *The Theory of Tikhonov Regularization for Fredholm Equations of the First Kind*, Pitman, London, 1984.

The error estimate for the conjugate gradient method in terms of the induced Q_α norm is taken from

J. T. King, A minimal error conjugate gradient method for ill-posed problems, *J. Optim. Theory Appl.* **60** (1989), 297–304.

2.4. Here we follow in part the paper by D. A. Murio and C. E. Mejía mentioned previously.

2.5. For an in-depth discussion of the difficulties associated with the accurate computation of matrix eigenvalues, the reader may consult—and enjoy—the book by

B. Parlett, *The Symmetric Eigenvalue Problem*, Prentice-Hall, Englewood Cliffs, NJ, 1980.

3

INVERSE HEAT
CONDUCTION PROBLEM

The classical heat conduction problem—where the heat flux or temperature histories at the surface of the body are known as functions of time and the interior temperature distribution is then determined—is termed a direct problem. However, in many dynamic heat transfer situations, it is necessary to estimate the surface heat flux and/or surface temperature histories from transient, measured temperatures inside a thermally conducting solid. This inverse heat conduction problem (IHCP) is frequently encountered, for example, in the determination of thermal constants in some quenching processes, the estimation of surface heat transfer measurements taken within the skin of a reentry vehicle, the determination of aerodynamic heating in wind tunnels and rocket nozzles, the design and development of calorimeter-type instrumentation, and infrared computerized tomography.

We begin with the investigation of the IHCP in its simplest geometry— associated with a one-dimensional infinite conductor—and then continue with the analysis of the one-dimensional IHCP corresponding to the finite slab case. The much more general and challenging situations related with the two-dimensional IHCP are studied next, in Chap. 4.

In what follows, we assume all the functions involved to be $L^2(-\infty, \infty)$ after extending them—if necessary—to the whole t axis by defining the functions to be 0 for $t < 0$.

3.1 ONE-DIMENSIONAL IHCP IN A SEMI-INFINITE BODY

Consider a semi-infinite slab with one-dimensional symmetry. Linear heat conduction with constant thermal properties is assumed and, after appropriate changes in the space and time scales, without loss of generality, the problem is normalized by using dimensionless quantities.

The problem can be described mathematically as follows.

The unknown temperature $u(x, t)$ satisfies

$$u_{xx} = u_t, \qquad 0 < x, t > 0, \tag{3.1a}$$

$$u(1, t) = F(t), \qquad t > 0, \tag{3.1b}$$

with corresponding approximate data function $F_m(t)$,

$$u(x, 0) = 0, \qquad 0 \le x, \tag{3.1c}$$

$$u(0, t) = f(t), \qquad t > 0, \text{ unknown}, \tag{3.1d}$$

$$u(x, t) \text{ bounded as } x \to \infty, t > 0, \tag{3.1e}$$

where t is time and x is the distance measured from the heated surface. We notice that the unknown boundary condition might be replaced by

$$-u_x(0, t) = q(t), \qquad t > 0, \text{ unknown}. \tag{3.1d$'$}$$

The objective is to estimate the surface temperature $f(t)$ and/or the surface heat flux $q(t)$ given the interior measurements at $x = 1$, denoted $F_m(t)$.

We assume that the measured data function $F_m(t)$ satisfies the $L^2(-\infty, \infty)$ error bound

$$\|F - F_m\| \le \varepsilon, \tag{3.2}$$

where ε is a known positive upper bound.

Taking the Fourier transform of (3.1a) with respect to time, we obtain the differential equation

$$\hat{u}_{xx}(x, w) = iw\hat{u}(x, w), \qquad 0 < x < \infty, -\infty < w < \infty,$$

which has the general solution

$$\hat{u}(x, w) = A(w) \exp\left[\sqrt{\frac{|w|}{2}}(1 + i\sigma)x\right] + B(w) \exp\left[-\sqrt{\frac{|w|}{2}}(1 + i\sigma)x\right],$$

$$i = \sqrt{-1}, \sigma = \text{sgn } w.$$

After imposing the boundary conditions (3.1d) and (3.1e), the general solution reduces to

$$\hat{u}(x, w) = B(w) \exp\left[-\sqrt{\frac{|w|}{2}}(1 + i\sigma)x\right],$$

and, in particular, for $x = 1$,

$$\hat{F}(w) = \hat{f}(w) \exp\left[-\sqrt{\frac{|w|}{2}} (1 + i\sigma)\right]$$

and

$$\hat{f}(w) = \hat{F}(w) \exp\left[\sqrt{\frac{|w|}{2}} (1 + i\sigma)\right]. \quad \blacksquare \qquad (3.3)$$

It is clear from (3.3) that the IHCP—attempting to go from $\hat{F}$ to $\hat{f}$—magnifies the high-frequency component by the factor $\exp[\sqrt{|w|}/2]$. Since we have assumed that f belongs to $L^2(-\infty, \infty)$, this implies that the exact data function F is such that its high-frequency components die out faster than $\exp[\sqrt{|w|}/2]$ for the product $\hat{F}\exp[\sqrt{|w|}/2] = \hat{f}$, and consequently f, to be in $L^2(-\infty, \infty)$. However, when we actually have to solve the inverse problem, instead of the ideal data function $F(t)$ we are given a noisy, measured data function $F_m(t)$ which is an $L^2(-\infty, \infty)$ function with its high-frequency components *not subject to such a rapidly decreasing behavior*, and we cannot, in general, guarantee that the product $\hat{F}_m \exp[\sqrt{|w|}/2]$ will be in $L^2(-\infty, \infty)$. Thus, we conclude that the IHCP is a highly ill-posed problem in the high-frequency Fourier components. Notice that for the IHCP the high-frequency components of the error grow exponentially, whereas for the differentiation problem they grow linearly; that is, the IHCP is much more ill posed—from this point of view—than the differentiation process.

Remark. In this chapter we always assume zero initial temperature for the IHCP. In fact, if $z(x, t)$ is the temperature distribution that satisfies the conditions stated in (3.1), with $z(x, 0) = h(x)$ instead of $z(x, 0) = 0$, we can first solve a direct problem with temperature function $v(x, t)$ that satisfies $v(0, t) = 0$ and $v(x, 0) = h(x)$. Considering the temperature function $u(x, t) = z(x, t) - v(x, t)$, it follows—by superposition—that $u(x, t)$ satisfies all the conditions (3.1). In particular, $u(x, 0) = 0$.

3.2   STABILIZED PROBLEMS

There are many ways to stabilize the IHCP. In this section we discuss and compare three different approaches to the approximate solution of the IHCP that successfully restore a certain type of continuous dependence on the data. All these procedures share the important property that they can be numerically implemented by means of suitable explicit finite-difference schemes marching in space and, consequently, can be applied to problems with nonconstant coefficients and to nonlinear problems. In all cases, we study the formal stability of the method and analyze the relationship between this property and the behavior of the high-frequency components of the noisy data. Other methods, based on the reformulation of the IHCP as a Volterra integral equation of the first kind, will be reviewed later in Sec. 3.5 when we present the numerical algorithms.

Our first method is the familiar mollification method (MM), introduced in Chap. 1 for the regularization of the numerical differentiation process.

Instead of attempting to find the point values of the temperature function $f(t)$, we lower our goal and attempt, instead, to reconstruct the δ-mollification of the function f at time t, given by $J_\delta f(t) = (\rho_\delta * f)(t)$. The stability analysis for this method follows now—as usual—from the definition of $J_\delta f$ and the mollification of (3.3). Indeed, we have

$$\widehat{(J_\delta f)}(w) = \frac{1}{\sqrt{2\pi}} \exp\left[\sqrt{\frac{|w|}{2}}(1 + i\sigma) - \frac{w^2\delta^2}{4}\right]\hat{F}(w). \tag{3.4}$$

If $\widehat{(J_\delta f_m)}(w)$ denotes the Fourier transform of the mollified surface temperature function $f(t)$ when the Fourier transform of the ideal data temperature $F(t)$ is replaced in (3.3) by the Fourier transform of the noisy data function $F_m(t)$, we obtain

$$\widehat{(J_\delta f_m)}(w) = \frac{1}{\sqrt{2\pi}} \exp\left[\sqrt{\frac{|w|}{2}}(1 + i\sigma) - \frac{w^2\delta^2}{4}\right]\hat{F}_m(w). \tag{3.5}$$

We observe that the function $J_\delta f_m \in L^2(-\infty, \infty)$ and

$$\sup_{0 < w < \infty} \left|\exp\left[\sqrt{\frac{|w|}{2}}(1 + i\sigma) - \frac{w^2\delta^2}{4}\right]\right| \le \exp\left[\frac{3}{4\delta^{2/3}}\right]. \tag{3.6}$$

Subtracting (3.5) from (3.4), squaring, integrating with respect to w, and using the estimate (3.6), we get

$$\left\|\widehat{(J_\delta f)} - \widehat{(J_\delta f_m)}\right\|^2 \le \frac{1}{2\pi} \exp\left[\frac{3}{2\delta^{2/3}}\right]\left\|\hat{F} - \hat{F}_m\right\|^2. \tag{3.7}$$

Thus, from Parseval's equality and assumption (3.2), it follows that

$$\|J_\delta f - J_\delta f_m\| \le \varepsilon \frac{1}{\sqrt{2\pi}} \exp\left[\frac{3}{4\delta^{2/3}}\right], \tag{3.8}$$

that is, the mollification method is formally stable. We have Lipschitz continuous dependence on the data as ε tends to 0, provided we keep the radius of mollification, δ, fixed. ∎

For the estimation of the error $\|f - J_\delta f_m\|$, we need to study the quantity $\|f - J_\delta f\|$—related with the consistency of the method—because in the absence of noise in the data, we require the mollified surface temperature to be "close" to the exact temperature solution. We know—recall Theorem A.9 —that $\|f - J_\delta f\| \to 0$ as $\delta \to 0$, but in order to estimate the rate of convergence, it is convenient to assume some extra smoothness on the ideal temperature solution function f. For instance, if $f' \in L^2(-\infty, \infty)$ and $\|f'\| \le M$, we immediately have

$$\|f - J_\delta f\| = \left\| \hat{f} - \widehat{(J_\delta f)} \right\| \tag{3.9}$$

$$= \left\| \hat{f}(w) \left(1 - \exp\left[-\frac{w^2 \delta^2}{4} \right] \right) \right\|$$

$$= \left\| iw\hat{f}(w) \frac{1 - \exp\left[-w^2\delta^2/4 \right]}{iw} \right\|$$

$$\le \|f'\| \sup_{-\infty < w < \infty} \left| \frac{1 - \exp\left[-w^2\delta^2/4 \right]}{w} \right|$$

$$\le M \sup_{-\infty < w < \infty} \left| \frac{1 - \exp\left[-w^2\delta^2/4 \right]}{w} \right|,$$

using Parseval's equality and recalling that $iw\hat{f}(w) = \hat{f}'(w)$.

Evaluating the right-hand side of expression (3.9) at the point $w \cong 2.2418\delta$ where the maximum is achieved, we get the consistency estimate

$$\|f - J_\delta f\| \le \tfrac{1}{2}M\delta. \quad ∎ \tag{3.10}$$

Combining this result with inequality (3.8), we obtain the following proposition.

Theorem 3.1 (Error Estimate)   Suppose that $\|F - F_m\| \le \varepsilon$. Then

(i) The problem

$$J_\delta u_{xx} = (J_\delta u)_t, \qquad 0 < x, t > 0,$$

$$J_\delta u(1, t) = J_\delta F_m(t), \qquad t > 0,$$

$$J_\delta u(x, 0) = 0, \qquad 0 \le x, \tag{3.11}$$

$$J_\delta u(0, t) = J_\delta f_m(t), \qquad t > 0, \text{ unknown},$$

$$J_\delta u(x, t) \text{ bounded as } x \to \infty, t > 0,$$

is a formally stable problem with respect to perturbations in the data.

(ii) If the exact boundary temperature $f(t)$ is such that $f' \in L^2(-\infty, \infty)$ with $\|f'\| \le M$, then $J_\delta f_m(t)$ verifies

$$\|f - J_\delta f_m\| \le \frac{1}{2} M\delta + \varepsilon \frac{1}{\sqrt{2\pi}} \exp\left[\frac{3}{4\delta^{2/3}}\right]. \qquad (3.12)$$

From a strictly theoretical point of view, with the parameter choice $\bar{\delta} = [\ln(1/\varepsilon)]^{-3/2}$, for example, we get convergence. Indeed, this particular choice—no "optimal" choice of the parameter is attempted—gives the error estimate, for small $\varepsilon > 0$,

$$\|f - J_{\bar{\delta}} f_m\| \le \frac{M}{2[\ln(1/\varepsilon)]^{3/2}} + \varepsilon^{1/4}.$$

The second method in this section—the hyperbolic method (HM)—is based on a completely different approach. The fundamental idea consists now on replacing the heat diffusion operator by a closely related hyperbolic one. From this point of view, the procedure can be classified as a singular perturbation method.

In an attempt to stabilize the IHCP, we might solve instead, the approximate hyperbolic problem

$$v_{xx} = \frac{1}{c^2} v_{tt} + v_t, \qquad 0 < x, t > 0, \qquad (3.13\text{a})$$

$$v(1, t) = F(t), \qquad t > 0, \qquad (3.13\text{b})$$

with corresponding approximate data function $F_m(t)$,

$$v(x, 0) = 0, \qquad 0 \le x, \qquad (3.13\text{c})$$

$$v_t(x, 0) = 0, \qquad 0 \le x, \qquad (3.13\text{d})$$

$$v(0, t) = f_c^H(t), \qquad t > 0, \text{ unknown}, \qquad (3.13\text{e})$$

$$v(x, t) \text{ bounded as } x \to \infty, t > 0. \qquad (3.13\text{f})$$

The parameter c, $c \gg 1$, can be interpreted as a finite thermal propagation speed. For most conduction problems its effect is negligible, although the traditional Fourier equation for heat flow implicitly assumes the thermal propagation speed to be infinite.

Taking the Fourier transform of (3.13a) with respect to time, we obtain the differential equation

$$\hat{v}_{xx}(x, w) = (iw - w^2/c^2)\hat{v}(x, w), \qquad 0 < x, -\infty < w < \infty,$$

which has the general solution

$$\hat{v}(x,w) = A(w) \exp\left[\sqrt{\frac{|w|}{2}} I(w,c)x\right] + B(w) \exp\left[-\sqrt{\frac{|w|}{2}} I(w,c)x\right],$$

where

$$I(w,c) = \left[\sqrt{1 + \frac{w^2}{c^4}} - \frac{|w|}{c^2}\right]^{1/2} + i\sigma \left[\sqrt{\frac{w^2}{c^4}} + \frac{|w|}{c^2}\right]^{1/2}.$$

Applying the boundary conditions (3.13e) and (3.13f), the general solution is reduced to

$$\hat{v}(x,w) = \left(\widehat{f_c^H}\right)(w) \exp\left[-\sqrt{\frac{|w|}{2}} I(w,c)x\right]$$

and, in particular, for $x = 1$,

$$\hat{F}(w) = \left(\widehat{f_c^H}\right)(w) \exp\left[-\sqrt{\frac{|w|}{2}} I(w,c)\right]$$

and

$$\left(\widehat{f_c^H}\right)(w) = \hat{F}(w) \exp\left[\sqrt{\frac{|w|}{2}} I(w,c)\right]. \tag{3.14}$$

For c fixed, $\left|\exp[\sqrt{|w|/2}\, I(w,c)]\right|$ is an increasing function of $|w|$ in $(0, \infty)$, bounded above. In fact,

$$\sup_{-\infty < w < \infty} \left|\exp\left[\sqrt{\frac{|w|}{2}} I(w,c)x\right]\right| \le \exp\left(\frac{c}{2}\right). \tag{3.15}$$

Notice that for low-frequency components, $I(w,c) \cong 1 + i\sigma$ and the hyperbolic model approximates the parabolic one very closely.

If $(\widehat{f_{c,m}^H})(w)$ denotes the Fourier transform of the hyperbolic surface temperature function when the Fourier transform of the ideal data temperature $F(t)$ is replaced in (3.14) by the Fourier transform of the noisy data function $F_m(t)$, we obtain

$$\left(\widehat{f_{c,m}^H}\right)(w) = \widehat{F_m}(w) \exp\left[\sqrt{\frac{|w|}{2}} I(w,c)\right]. \tag{3.16}$$

Once again, using inequality (3.15), we see that the function $f_{c,m}^H \in L^2(-\infty, \infty)$ since $F_m \in L^2(-\infty, \infty)$. Subtracting (3.16) from (3.14), taking norms, and using inequality (3.15) and Parseval's equality, it follows that

$$\|f_c^H - f_{c,m}^H\| \le \varepsilon \exp\left(\frac{c}{2}\right). \quad \blacksquare \qquad (3.17)$$

This shows that the hyperbolic method is also formally stable. We have restored continuous dependence on the data as ε tends to 0, provided that we keep the finite thermal propagation speed, c, fixed.

If $f' \in L^2(-\infty, \infty)$ and $M = \max(\|f\|, \|f'\|)$, it is possible to show—see the references at the end of this chapter—that

$$\|f - f_c^H\| \le \frac{2}{\sqrt{3}} \frac{M}{c^{2/3}}. \qquad (3.18)$$

This consistency estimate together with inequality (3.17) gives the following error estimate.

Theorem 3.2 (Error Estimate)   Suppose that $\|F - F_m\| \le \varepsilon$. Then

    (i) Problem (3.13) is a formally stable problem with respect to perturbations in the data.

    (ii) If the exact boundary temperature $f(t)$ is such that $f' \in L^2(-\infty, \infty)$ with $\max(\|f\|, \|f'\|) \le M$, then $f_{c,m}^H(t)$ verifies

$$\|f - f_{c,m}^H\| \le \varepsilon \exp\left(\frac{c}{2}\right) + \frac{2}{\sqrt{3}} \frac{M}{c^{2/3}}. \qquad (3.19)$$

As with the mollification method, theoretically, with the choice $\bar{c} = 2\ln(1/\sqrt{\varepsilon})$, for example, we obtain convergence as $\varepsilon \to 0$. For small ε, evaluating the upper bound (3.19) at $\bar{c}$, we have

$$\|f - f_{\bar{c},m}^H\| \le 2^{1/3}M\left[\ln\left(\frac{1}{\sqrt{\varepsilon}}\right)\right]^{2/3} + \sqrt{\varepsilon}.$$

Remark   A closer look at the formal stability analysis of both the mollification and the hyperbolic methods reveals that, for the former, according to (3.5),

$$(\widehat{J_\delta f_m})(w) = \frac{1}{\sqrt{2\pi}} \exp\left[\sqrt{\frac{|w|}{2}}(1 + i\sigma) - \frac{w^2\delta^2}{4}\right]\hat{F}_m(w),$$

showing that the high-frequency components of $\hat{F}_m$ are effectively filtered; the kernel function

$$\exp\left[\sqrt{\frac{|w|}{2}}\,(1 + i\sigma) - \frac{w^2\delta^2}{4}\right] \to 0 \qquad \text{as} \quad |w| \to \infty$$

belongs to $L^2(-\infty, \infty)$ and the stability bound (3.8) is not sharp at all for high-frequency components. On the other hand, for the hyperbolic method, from (3.16), we have

$$\left(\widehat{f_{c,m}^H}\right)(w) = \widehat{F_m}(w)\exp\left[\sqrt{\frac{|w|}{2}}\,I(w, c)\right]$$

and

$$\exp\left[\sqrt{\frac{|w|}{2}}\,I(w, c)\right] \to \exp\left(\frac{c}{2}\right) \qquad \text{as} \quad |w| \to \infty.$$

The kernel function now allows all the frequencies to enter. The stability estimate (3.17) is now sharp for high-frequency components; they are amplified by the factor $\cong \exp(c/2)$.

The third method in this section is motivated from the realization that it is possible to design a procedure "in between" the two previous methods. This method—appropriately named the intermediate method (IM)—should be stable, barely filter (not necessarily eliminate), and certainly not amplify the high-frequency components of the measured data function. Exploiting the powerful idea of singular perturbation, we propose to replace the heat diffusion equation by the approximating partial differential equation

$$z_{xx} = z_t - \gamma^2 z_{ttx}, \qquad 0 < x, t > 0, \qquad (3.20a)$$

$$z(1, t) = F(t), \qquad\qquad t > 0, \qquad (3.20b)$$

with corresponding approximate data function $F_m(t)$,

$$z(x, 0) = 0, \qquad\qquad 0 \le x, \qquad (3.20c)$$

$$z(0, t) = f_\gamma^I(t), \qquad t > 0, \text{ unknown} \qquad (3.20d)$$

$$z(x, t) \text{ bounded as } x \to \infty, t > 0, \qquad (3.20e)$$

$$z(x, t) \text{ bounded as } t \to \infty, x > 0. \qquad (3.20f)$$

Condition (3.20f) is imposed to ensure the well-posedness of system (3.20) considered as a Cauchy problem in the time variable with $z(0, t)$ known and condition (3.20b) deleted.

In dimensionless quantities, if $q(x, t)$ and $\tau(x, t)$ represent, respectively, the heat flux per unit area and the temperature in a solid, the first law of thermodynamics for the one-dimensional heat flow is given by

$$q_x + \tau_t = 0, \tag{3.21}$$

and the Fourier equation of thermal conductivity for heat flow states mathematically that

$$q + \tau_x = 0. \tag{3.22}$$

The contribution of these two equations forms the governing parabolic partial differential equation for transient heat conduction (3.1a).

Physically, the model equation (3.20a) can be obtained by combining (3.21) (with $\tau = z$) and the hypothetical—experimentally not established—Fourier law

$$\gamma^2 z_{tt} + q + z_x = 0, \tag{3.23}$$

where γ, $\gamma \ll 1$, is a nonnegative constant and γ^2 can be interpreted as a relaxation time parameter.

Taking the Fourier transform of (3.20a) with respect to time, we obtain the differential equation

$$\hat{z}_{xx}(x, w) - w^2 \gamma^2 \hat{z}_x(x, w) - iw\hat{z}(x, w) = 0, \qquad 0 < x, \ -\infty < w < \infty,$$

whose general solution is given by

$$\hat{z}(x, w) = A(w) \exp\left[\left\{\frac{1}{2}\gamma^2 w^2 + K(w, \gamma)\right\}x\right]$$
$$+ B(w) \exp\left[\left\{\frac{1}{2}\gamma^2 w^2 + K(w, \gamma)\right\}x\right],$$

where

$$K(w, \gamma) = 2^{-3/2}\left\{\left[(\gamma^8 w^8 + 16w^2)^{1/2} + \gamma^4 w^4\right]^{1/2}\right.$$
$$\left. + i\sigma\left[(\gamma^8 w^8 + 16w^2)^{1/2} - \gamma^4 w^4\right]^{1/2}\right\}.$$

Applying the boundary conditions (3.20d) and (3.20e), the general solution becomes

$$\hat{z}(x,w) = \left(\widehat{f_\gamma^I}\right)(w) \exp\left[\left\{\frac{1}{2}\gamma^2 w^2 - K(w,\gamma)\right\}x\right]$$

and, in particular, for $x = 1$,

$$\hat{F}(w) = \left(\widehat{f_\gamma^I}\right)(w) \exp\left[\frac{1}{2}\gamma^2 w^2 - K(w,\gamma)\right]$$

and

$$\left(\widehat{f_\gamma^I}\right)(w) = \hat{F}(w) \exp\left[-\frac{1}{2}\gamma^2 w^2 + K(w,\gamma)\right]. \qquad (3.24)$$

For each γ, the function $G_\gamma(w) = |\exp[-\frac{1}{2}\gamma^2 w^2 + K(w,\gamma)]|$ attains its maximum at the point

$$|w_0| = \left(\frac{9\sqrt{17} - 37}{4\gamma^8}\right)^{1/6} \cong 0.54768\gamma^{-4/3} \qquad (3.25)$$

and

$$\lim_{|w| \to \infty} G_\gamma(w) = 1. \qquad (3.26)$$

Furthermore,

$$\sup_{-\infty < w < \infty} \left|\exp\left[-\frac{1}{2}\gamma^2 w^2 + K(w,\gamma)\right]\right|$$

$$= \left|\exp\left[-\frac{1}{2}\gamma^2 w_0^2 + K(w_0,\gamma)\right]\right|$$

$$\le \exp\left(\frac{1}{2}\gamma^{-2/3}\right). \qquad (3.27)$$

Notice that for low-frequency components, the intermediate method behaves like the original parabolic problem since

$$\exp[K(w,\gamma)] \cong \exp\left[\sqrt{\frac{|w|}{2}}(1 + i\sigma)\right].$$

If $(\widehat{f_{\gamma,m}^I})(w)$ denotes the Fourier transform of the unknown surface temperature for the intermediate method when the Fourier transform of the ideal data temperature $F(t)$ is replaced in (3.24) by the Fourier transform of the noisy data function $F_m(t)$, we have

$$\left(\widehat{f^I_{\gamma, m}}\right)(w) = \hat{F}_m(w) \exp\left[-\frac{1}{2}\gamma^2 w^2 + K(w, \gamma)\right]. \tag{3.28}$$

It follows immediately, using inequality (3.27), that the function $f^I_{\gamma, m} \in L^2(-\infty, \infty)$. From (3.28) and (3.24) it follows, in the usual manner, that

$$\|f^I_\gamma - f^I_{\gamma, m}\| \le \varepsilon \exp\left(\frac{1}{2}\gamma^{-2/3}\right). \quad \blacksquare \tag{3.29}$$

This shows that the new singular perturbation method is formally stable. We have restored continuous dependence on the data as ε tends to 0 provided we keep the time relaxation parameter, γ, fixed.

We observe that the stability bound (3.29) for the intermediate method is the exact counterpart of the one corresponding to the mollification method after interchanging the parameters δ and γ [see (3.8)]. Also, we notice that for the new method, the kernel function

$$\exp\left[-\frac{1}{2}\gamma^2 w^2 + K(w, \gamma)\right] \to 1 \quad \text{as} \quad |w| \to \infty,$$

and the stability bound (3.29) is not sharp for the high-frequency components. In fact, these frequencies are barely amplified, according to (3.26). In this respect, the intermediate method is "inserted" between the mollification method—which eliminates the high-frequency components of the noisy data function—and the hyperbolic method—which amplifies those frequencies by a factor $\cong \exp(c/2)$ We shall see in Sec. 3.6 that the particular control of the high-frequency components introduced by the intermediate method is sufficient for the successful implementation of a conditionally stable, fully explicit, finite-difference space marching scheme.

If $f' \in L^2(-\infty, \infty)$ and $M = \max(\|f\|, \|f'\|)$, it is possible to show—see the references at the end of this chapter—that

$$\|f^I_\gamma - f\| \le \left(\gamma + 2\sqrt{\gamma}\right)M = O\left(\sqrt{\gamma}\right). \tag{3.30}$$

Theoretically, the choice $\bar{\gamma} = [\ln(1/\varepsilon)]^{-3/2}$, for example, implies convergence as $\varepsilon \to 0$. From (3.29) and (3.30) plus the triangle inequality, we have

$$\|f - f^I_{\bar{\gamma}, m}\| \le 3\left[\ln\left(\frac{1}{\varepsilon}\right)\right]^{-3/4} M + \sqrt{\varepsilon}. \quad \blacksquare \tag{3.31}$$

For completeness, we state the following proposition.

Theorem 3.3 (Error Estimate)   Suppose that $\|F - F_m\| \le \varepsilon$. Then

(i) Problem (3.20) is a formally stable problem with respect to perturbations in the data.
(ii) If the exact boundary temperature $f(t)$ is such that $f' \in L^2(-\infty, \infty)$ with $\max(\|f\|, \|f'\|) \le M$, then f_γ^I verifies inequality (3.31).

3.3 ONE-DIMENSIONAL IHCP WITH FINITE SLAB SYMMETRY

The most striking and challenging difference in going from the IHCP in the semi-infinite slab to the IHCP in the finite slab is that now we have to deal with a system of two equations with two unknowns instead of just one equation with one unknown.

We consider a one-dimensional IHCP in which the temperature and the heat flux histories $f(t)$ and $q(t)$ on the left surface $x = 0$ are desired and unknown, and the temperature and heat flux histories $F(t)$ and $Q(t)$, respectively, are approximately measured at the right surface $x = 1$. In dimensionless quantities, the normalized linear IHCP can be described mathematically as follows:

The unknown temperature $u(x, t)$ satisfies

$$u_{xx} = u_t, \qquad 0 < x < 1, t > 0, \qquad (3.32a)$$

$$u(1, t) = F(t), \qquad t > 0, \qquad (3.32b)$$

with corresponding approximate data function $F_m(t)$,

$$u_x(1, t) = Q(t), \qquad t > 0, \qquad (3.32c)$$

with corresponding approximate data function $Q_m(t)$,

$$u(0, t) = f(t), \qquad t > 0, \qquad (3.32d)$$

the desired but unknown temperature function,

$$u_x(0, t) = q(t), \qquad t > 0, \qquad (3.32e)$$

the desired but unknown heat flux function,

$$u(x, 0) = 0, \qquad 0 \le x \le 1. \qquad (3.32f)$$

It is now quite natural to assume that the exact extended data functions $F(t)$ and $Q(t)$ and the extended measured data functions $F_m(t)$ and $Q_m(t)$ satisfy the L^2 error bounds

$$\|F - F_m\| \le \varepsilon \qquad \text{and} \qquad \|Q - Q_m\| \le \varepsilon. \tag{3.33}$$

The classical—well-posed—direct problems associated with the finite slab geometry are of course obtained by specifying *one* of the functions F or Q at the surface $x = 1$ and *one* of the functions f or q at the surface $x = 0$.

Applying a Fourier transformation with respect to time to (3.32a), we obtain

$$\hat{u}_{xx}(x, w) = iw\hat{u}(x, w), \qquad 0 < x < 1, \ -\infty < w < \infty. \tag{3.34}$$

Expression (3.34) is a second-order differential equation which has the general solution

$$\hat{u}(x, w) = A(w) \exp\left[\sqrt{\frac{|w|}{2}}\,(1 + i\sigma)x\right] + B(w) \exp\left[-\sqrt{\frac{|w|}{2}}\,(1 + i\sigma)x\right], \tag{3.35}$$

where $\sigma = \operatorname{sgn} w$.

With $\alpha = \sqrt{|w|/2}$ and $\beta = 1 + i\sigma$, differentiation of (3.35) with respect to x gives

$$\hat{u}_x(x, w) = \alpha\beta A e^{\alpha\beta x} - \alpha\beta B e^{-\alpha\beta x}. \tag{3.36}$$

When $x = 0$, it follows from expressions (3.35) and (3.36), using conditions (3.32d) and (3.32e), that

$$\hat{f}(w) = A(w) + B(w)$$

and

$$\hat{q}(w) = \alpha\beta(A(w) - B(w)).$$

Solving this system for the functions $A(w)$ and $B(w)$, we get

$$\begin{pmatrix} A(w) \\ B(w) \end{pmatrix} = \frac{1}{2\alpha\beta} \begin{pmatrix} \alpha\beta & 1 \\ \alpha\beta & -1 \end{pmatrix} \begin{pmatrix} \hat{f}(w) \\ \hat{q}(w) \end{pmatrix}. \tag{3.37}$$

On the other hand, according to (3.35) and (3.36), for $x = 1$, using conditions (3.32b) and (3.32c), we have

$$\begin{pmatrix} \hat{F}(w) \\ \hat{Q}(w) \end{pmatrix} = \begin{pmatrix} e^{\alpha\beta} & e^{-\alpha\beta} \\ \alpha\beta e^{\alpha\beta} & -\alpha\beta e^{-\alpha\beta} \end{pmatrix} \begin{pmatrix} A(w) \\ B(w) \end{pmatrix}.$$

Combining this last equality with (3.37), it follows that

$$
\begin{pmatrix} \hat{F}(w) \\ \hat{Q}(w) \end{pmatrix} = \frac{1}{2\alpha\beta} \begin{pmatrix} e^{\alpha\beta} & e^{-\alpha\beta} \\ \alpha\beta e^{\alpha\beta} & -\alpha\beta e^{-\alpha\beta} \end{pmatrix} \begin{pmatrix} \alpha\beta & 1 \\ \alpha\beta & -1 \end{pmatrix} \begin{pmatrix} \hat{f}(w) \\ \hat{q}(w) \end{pmatrix}.
$$

Thus,

$$
\begin{pmatrix} \hat{F}(w) \\ \hat{Q}(w) \end{pmatrix} = \begin{pmatrix} \cosh\alpha\beta & \dfrac{1}{\alpha\beta}\sinh\alpha\beta \\ \alpha\beta\sinh\alpha\beta & \cosh\alpha\beta \end{pmatrix} \begin{pmatrix} \hat{f}(w) \\ \hat{q}(w) \end{pmatrix} \tag{3.38}
$$

and

$$
\begin{pmatrix} \hat{f}(w) \\ \hat{q}(w) \end{pmatrix} = \begin{pmatrix} \cosh\alpha\beta & -\dfrac{1}{\alpha\beta}\sinh\alpha\beta \\ -\alpha\beta\sinh\alpha\beta & \cosh\alpha\beta \end{pmatrix} \begin{pmatrix} \hat{F}(w) \\ \hat{Q}(w) \end{pmatrix}. \quad \blacksquare \tag{3.39}
$$

The transformations (3.38) and (3.39) clearly illustrate the symmetry of the IHCP in the finite slab. Moreover, since $|\sinh\alpha\beta| = |\sinh(\sqrt{|w|}/2\,(1 + i\sigma))|$ $\to \infty$ as $|w| \to \infty$ and $|\cosh\alpha\beta| = |\cosh(\sqrt{|w/2}\,(1 + i\sigma))| \to \infty$ as $|w| \to \infty$, we see that solving the inverse problem (3.32)—obtaining $f(t)$ and $q(t)$ from $F(t)$ and $Q(t)$—amplifies the error in a high-frequency component by the factor $\exp(\sqrt{|w|/2}\,)$, showing that the IHCP is highly ill posed in the high-frequency components.

The mollification method stabilizes this problem by attempting to reconstruct the δ-mollifications $J_\delta f$ and $J_\delta q$ of the functions f and q, respectively, at time t.

Mollifying system (3.32) with respect to t, we obtain the following associated problem.

Attempt to find $J_\delta f_m(t) = J_\delta u(0, t)$ and $J_\delta q_m(t) = J_\delta u_x(0, t)$ at some time t of interest and for some radius $\delta > 0$, given that $J_\delta u(x, t)$ satisfies

$$
\begin{aligned}
J_\delta u_{xx} &= (J_\delta u)_t, & 0 < x < 1,\, t > 0, \\[4pt]
J_\delta u(1, t) &= J_\delta F_m(t), & t > 0, \\[4pt]
J_\delta u_x(1, t) &= J_\delta Q_m(t), & t > 0, \\[4pt]
J_\delta u(x, 0) &= 0, & 0 \le x \le 1.
\end{aligned} \tag{3.40}
$$

This problem and its solutions satisfy the following proposition.

Theorem 3.4 (Error Estimate)   Assume that $\|F - F_m\| \leq \varepsilon$ and $\|Q - Q_m\| \leq \varepsilon$. Then

(i) Problem (3.40) is a formally stable problem with respect to perturbations in the data.

(ii) If the exact boundary temperature $f(t)$ and heat flux $q(t)$ have uniformly bounded first derivatives on some finite interval $I = [0, T]$, that is, $\max(\|f'\|_I, \|q'\|_I) \leq M$, then $J_\delta f_m(t)$ and $J_\delta q_m(t)$ verify

$$\|f - J_\delta f_m\|_I \leq \frac{1}{2}M\delta + \varepsilon \exp\left[\frac{1}{(2\delta)^{2/3}}\right]$$

and

$$\|q - J_\delta q_m\|_I \leq \frac{1}{2}M\delta + 2\varepsilon \exp\left[\frac{1}{(2\delta)^{2/3}}\right].$$

Proof

(i) Repeating the same procedure as the one used to obtain (3.39) and taking into account the definitions of $J_\delta f$ and $J_\delta q$, we have

$$\left\|\widehat{(J_\delta f)} - \widehat{(J_\delta f_m)}\right\| \leq \left\|\frac{1}{\sqrt{2\pi}}e^{-w^2\delta^2/4}\cosh\alpha\beta\left(\hat{F} - \hat{F}_m\right)\right\|$$

$$+ \left\|\frac{1}{\sqrt{2\pi}}e^{-w^2\delta^2/4}\frac{1}{\alpha\beta}\sinh\alpha\beta\left(\hat{Q} - \hat{Q}_m\right)\right\| \quad (3.41)$$

and

$$\left\|\widehat{(J_\delta q)} - \widehat{(J_\delta q_m)}\right\| \leq \left\|\frac{1}{\sqrt{2\pi}}\alpha\beta e^{-w^2\delta^2/4}\sinh\alpha\beta\left(\hat{F} - \hat{F}_m\right)\right\|$$

$$+ \left\|\frac{1}{\sqrt{2\pi}}e^{-w^2\delta^2/4}\cosh\alpha\beta\left(\hat{Q} - \hat{Q}_m\right)\right\|. \quad (3.42)$$

We notice that

$$|\cosh\alpha\beta| \leq \frac{1}{2}\exp\sqrt{\frac{|w|}{2}}\left(1 + \exp\left[-\sqrt{\frac{|w|}{2}}\right]\right),$$

$$|\alpha\beta\sinh\alpha\beta| \leq \frac{1}{2}\sqrt{|w|}\left(\exp\left[2\sqrt{\frac{|w|}{2}}\right] + 1\right),$$

and

$$\left| \frac{1}{\alpha\beta} \sinh \alpha\beta \right| \le \left\{ \frac{1}{\sqrt{2|w|}} \left(\cosh \sqrt{2|w|} - \cos \sqrt{2|w|} \right) \right\}^{1/2}$$

$$\le \left\{ 2 \sum_{k=0}^{\infty} \frac{\left(\sqrt{2|w|} \right)^{2k}}{(2(k+1))!} \right\}^{1/2} \le \sqrt{2} \exp\sqrt{\frac{|w|}{2}} .$$

Replacing these estimates in (3.41), we get

$$\left\| \widehat{(J_\delta f)} - \widehat{(J_\delta f_m)} \right\| \le \frac{1}{\sqrt{2\pi}} \max_{|w| \ge 0} \left\{ \exp\left[\sqrt{\frac{|w|}{2}} - \frac{w^2 \delta^2}{4} \right] \right\} \| \hat{F} - \hat{F}_m \|$$

$$+ \frac{1}{\sqrt{\pi}} \max_{|w| \ge 0} \left\{ \exp\left[\sqrt{\frac{|w|}{2}} - \frac{w^2 \delta^2}{4} \right] \right\} \| \hat{Q} - \hat{Q}_m \|$$

$$\le \frac{2}{\sqrt{2\pi}} \varepsilon \exp\frac{3}{4(2\delta)^{2/3}} \le \varepsilon \exp\frac{1}{(2\delta)^{2/3}} ,$$

and by Parseval's equality, we obtain

$$\| J_\delta f - J_\delta f_m \| \le \varepsilon \exp\frac{1}{(2\delta)^{2/3}} . \tag{3.43}$$

Similarly, using the previous estimates in (3.42), we have

$$\left\| \widehat{(J_\delta q)} - \widehat{(J_\delta q_m)} \right\| \le \frac{\varepsilon}{\sqrt{2\pi}} \max_{|w| \ge 0} \left\{ \frac{1}{2} \sqrt{|w|} \exp\left[1 + 2\sqrt{\frac{|w|}{2}} - \frac{w^2 \delta^2}{4} \right] \right\}$$

$$+ \frac{\varepsilon}{\sqrt{2\pi}} \max_{|w| \ge 0} \left\{ \exp\left[\sqrt{\frac{|w|}{2}} - \frac{w^2 \delta^2}{4} \right] \right\}$$

$$\le \frac{\varepsilon}{\sqrt{2\pi}} \max_{|w| \ge 0} \left\{ \exp\left[1 + \sqrt{2|w|} - \frac{w^2 \delta^2}{4} \right] \right\}$$

$$+ \frac{\varepsilon}{\sqrt{2\pi}} \max_{|w| \ge 0} \left\{ \exp\left[\sqrt{\frac{|w|}{2}} - \frac{w^2 \delta^2}{4} \right] \right\}$$

$$\le \frac{\varepsilon}{\sqrt{2\pi}} \exp\left[1 + \frac{3}{4(2\delta)^{2/3}} \right]$$

$$+ \frac{\varepsilon}{\sqrt{2\pi}} \exp\left[\frac{3}{4(2\delta)^{2/3}} \right]$$

$$\le 2\varepsilon \exp\frac{1}{(2\delta)^{2/3}} ,$$

and by Parseval's equality,

$$\|J_\delta q - J_\delta q_m\| \le 2\varepsilon \exp \frac{1}{(2\delta)^{2/3}}. \tag{3.44}$$

Inequalities (3.43) and (3.44) indicate that for a fixed $\delta > 0$, the errors tend to 0 as $\varepsilon \to 0$. This shows that the mollified inverse problem (3.40) is formally stable with respect to perturbations in the data.

(ii) Write

$$f(t) - J_\delta f_m(t) = f(t) - J_\delta f(t) + J_\delta f(t) - J_\delta f_m(t)$$

and

$$q(t) - J_\delta q_m(t) = q(t) - J_\delta q(t) + J_\delta q(t) - J_\delta q_m(t).$$

Taking norms, using the estimates (3.10), (3.43), (3.44), and the triangle inequality, we have

$$\|J_\delta f - J_\delta f_m\|_I \le \frac{1}{2}M\delta + \varepsilon \exp \frac{1}{(2\delta)^{2/3}},$$

$$\|J_\delta q - J_\delta q_m\|_I \le \frac{1}{2}M\delta + 2\varepsilon \exp \frac{1}{(2\delta)^{2/3}}. \quad \blacksquare \tag{3.45}$$

The space marching finite-difference implementation of the methods introduced in the last two sections is described next.

3.4   FINITE-DIFFERENCE APPROXIMATIONS

Finite Slab Geometry

Mollification Method
In this section we consider the approximate solution of the mollified problem (3.40) by means of finite differences in detail.

Without loss of generality, we seek to reconstruct the unknown boundary temperature $J_\delta f_m$ and heat flux $J_\delta q_m$ in the time interval $I = [0, 1]$. We consider a uniform grid in the (x, t) plane: $\{(x_s = sh, t_j = jk), s = 0, 1, \ldots, N; Nh = 1; j = 0, 1, \ldots, M; Mk = L\}$, where L depends on h and k in a way to be specified later, $L > 1$.

We first notice that the differential equation $J_\delta u_{xx} = (J_\delta u)_t$ is being treated as a second-order ordinary differential equation when marching—

backward—in space. This initial value problem—technically a Cauchy problem in the setting of partial differential equations—needs two initial conditions at $x = 1$, given by the approximate knowledge of the mollified temperature data function $J_\delta u(1, t) = J_\delta F_m(t)$ and the mollified heat flux $J_\delta u_x(1, t) = J_\delta Q_m(t)$.

Thus, we consider the mollified problem

$$
\begin{aligned}
J_\delta u_{xx} &= (J_\delta u)_t, & 0 < x < 1, t > 0, \\
J_\delta u(1, t) &= J_\delta F_m(t), & t > 0, \\
J_\delta u_x(1, t) &= J_\delta Q_m(t), & t > 0, \\
J_\delta u(x, 0) &= 0, & 0 \le x \le 1, \\
J_\delta u(0, t) &= J_\delta f_m(t), & \text{unknown, } t > 0, \\
J_\delta u_x(0, t) &= J_\delta q_m(t), & \text{unknown, } t > 0.
\end{aligned} \tag{3.46}
$$

Introducing the new variables $v = J_\delta u$ and $w = \partial v / \partial x$, system (3.46) is equivalent to

$$
\begin{aligned}
\frac{\partial w}{\partial x} &= \frac{\partial v}{\partial t}, & 0 < x < 1, 0 < t, \\
w &= \frac{\partial v}{\partial x}, & 0 < x < 1, 0 < t, \\
v(1, t) &= J_\delta F_m, & 0 < t, \\
w(1, t) &= J_\delta Q_m, & 0 < t, \\
v(x, 0) &= 0, & 0 \le x \le 1, \\
v(0, t) &= J_\delta f_m, & 0 < t, \text{unknown}, \\
w(1, t) &= J_\delta q_m, & 0 < t, \text{unknown}.
\end{aligned} \tag{3.47}
$$

With the grid functions V and W respectively defined by

$$
V_s^j = v(x_s, t_j) \quad \text{and} \quad W_s^j = w(x_s, t_j), \qquad 0 \le s \le N, 0 \le j \le M,
$$

we notice that

$$
V_N^j = J_\delta F_m(t_j), \qquad W_N^j = J_\delta Q_m(t_j), \qquad 0 \le j \le M,
$$

and

$$
V_s^0 = 0, \qquad 0 \le s \le N.
$$

Approximating the partial differential equations in (3.46) with the consistent finite-difference schemes

$$\frac{W_s^j - W_{s-1}^j}{h} = \frac{V_s^{j+1} - V_s^{j-1}}{2k}$$

and

$$W_{s-1}^j = \frac{V_s^j - V_{s-1}^j}{h}, \qquad N \geq s \geq 1, 1 \leq j \leq M - 1,$$

we obtain a local truncation error that behaves like $O(h^2 + k^2)$ for the discrete mollified temperature and $O(h + k^2)$ for the discrete mollified heat flux as $h, k \to 0$. The discrete system associated with (3.47) is now given by

$$W_{s-1}^j = W_s^j - \frac{h}{2k}(V_s^{j+1} - V_s^{j-1}),$$

$$V_{s-1}^j = V_s^j - hW_{s-1}^j,$$

$$s = N, N - 1, \ldots, 1, j = 1, 2, \ldots, M - N + s - 1,$$

$$V_N^j = J_\delta F_m(jk), \qquad j = 1, 2, \ldots, M, \qquad (3.48)$$

$$W_N^j = J_\delta Q_m(jk), \qquad j = 1, 2, \ldots, M,$$

$$V_s^0 = 0, \qquad s = 0, 1, \ldots, N.$$

Notice that, as we march backward in space, at each step we must drop the estimation of the interior temperature from the highest previous point in time. Since we want to evaluate $\{V_0^j\}$ and $\{W_0^j\}$ at the grid points of $I = [0, 1]$ after N iterations, the minimum initial length of the data sample interval, $L = Mk$, has to be decided. That is, L needs to satisfy the condition $Mk - Nk = 1$. Hence, $L = Mk = 1 + k/h$.

Next, in order to analyze the numerical stability of system (3.48), assuming that the equations hold in the entire discrete plane (x, t), after applying the discrete Fourier transform—recall Definition A.16—system (3.48) becomes

$$\tilde{W}_{s-1}(w) = \tilde{W}_s(w) - i\frac{h}{k}\sin(kw)\tilde{V}_s(w),$$

$$\tilde{V}_{s-1}(w) = \tilde{V}_s(w) - h\tilde{W}_{s-1}(w),$$

$$\tilde{V}_N(w) = \widetilde{(J_\delta F_m)}(w), \qquad \text{given}, \qquad (3.49)$$

$$\tilde{W}_N(w) = \widetilde{(J_\delta Q_m)}(w), \qquad \text{given},$$

$$s = N, N - 1, \ldots, 1, 0 \leq |w| \leq \frac{\pi}{k}.$$

Equivalently,

$$\begin{pmatrix} \tilde{V}_{s-1} \\ \tilde{W}_{s-1} \end{pmatrix} = \begin{pmatrix} 1 + i\dfrac{h^2}{k}\sin(wk) & -h \\ -i\dfrac{h}{k}\sin(wk) & 1 \end{pmatrix} \begin{pmatrix} \tilde{V}_s \\ \tilde{W}_s \end{pmatrix},$$

$$s = N, N-1, \ldots, 1, \, 0 \le |w| \le \frac{\pi}{k}.$$

Hence,

$$|\tilde{V}_{s-1}| \le \left(1 + h + \frac{h^2}{k}\sin(|w|k)\right)\max\left(|\tilde{V}_s|, |\tilde{W}_s|\right)$$

and

$$|\tilde{W}_{s-1}| \le \left(1 + \frac{h}{k}\sin(|w|k)\right)\max\left(|\tilde{V}_s|, |\tilde{W}_s|\right).$$

Thus,

$$\max\left(|\tilde{V}_{s-1}|, |\tilde{W}_{s-1}|\right) \le \left(1 + h + \frac{h}{k}\sin(|w|k)\right)\max\left(|\tilde{V}_s|, |\tilde{W}_s|\right)$$

and

$$\max\left(|\tilde{V}_0|, |\tilde{W}_0|\right) \le \left(1 + h + \frac{h}{k}\sin(|w|k)\right)^N \max\left(|\tilde{V}_N|, |\tilde{W}_N|\right)$$

$$\le \left(1 + h + \frac{h}{k}\sin(|w|k)\right)^N \max\left(|\tilde{\rho}_\delta \tilde{F}_m|, |\tilde{\rho}_\delta \tilde{Q}_m|\right).$$

Consequently, we can write

$$|\tilde{V}_0|^2 \le \left\{\left(1 + h + \frac{h}{k}\sin(|w|k)\right)^N\right\}^2 |\tilde{\rho}_\delta|^2 \left(|\tilde{F}_m|^2 + |\tilde{Q}_m|^2\right) \quad (3.50a)$$

and

$$|\tilde{W}_0|^2 \le \left\{\left(1 + h + \frac{h}{k}\sin(|w|k)\right)^N\right\}^2 |\tilde{\rho}_\delta|^2 \left(|\tilde{F}_m|^2 + |\tilde{Q}_m|^2\right). \quad (3.50b)$$

Recalling that the relationship between the continuous Fourier transform $\hat{\rho}_\delta$ and the discrete Fourier transform $\tilde{\rho}_\delta$ is given by Poisson's summation formula

$$\tilde{\rho}_\delta(w) = \sum_{j=-\infty}^{\infty} \hat{\rho}_\delta\left(w + j\frac{2\pi}{k}\right), \qquad 0 \le |w| \le \frac{\pi}{k},$$

we have

$$
\begin{aligned}
|\tilde{\rho}_\delta(w)| &\le \frac{1}{\sqrt{2\pi}} \sum_{j=-\infty}^{\infty} \exp\left[-\frac{(w + 2j\pi/k)^2\delta^2}{4}\right] \\
&= \frac{1}{\sqrt{2\pi}} \exp\left[-\frac{w^2\delta^2}{4}\right] \sum_{j=-\infty}^{\infty} \exp\left[-\frac{(wkj\pi + j^2\pi^2)\delta^2}{k^2}\right] \\
&\le \frac{1}{\sqrt{2\pi}} \exp\left[-\frac{w^2\delta^2}{4}\right]\left(1 + 2\sum_{j=1}^{\infty} \exp\left[-\frac{\{(j-1)j\pi^2\delta^2\}}{k^2}\right]\right) \\
&\le \frac{1}{\sqrt{2\pi}} \exp\left[-\frac{w^2\delta^2}{4}\right]\left\{1 + 2\frac{1}{1 - \exp(-\pi^2\delta^2/k^2)}\right\} \\
&\le \frac{1}{\sqrt{2\pi}} \exp\left[-\frac{w^2\delta^2}{4}\right]\left\{1 + 2\frac{1}{1 - \exp(-\pi^2)}\right\} \\
&\le 4\exp\left[-\frac{w^2\delta^2}{4}\right],
\end{aligned}
$$

after utilizing the inequality $\delta \ge k$.

Using this last estimate in (3.50a) and (3.50b), integrating with respect to w, and taking square roots, we get

$$
\begin{aligned}
\max\left(\|\tilde{V}_0\|_2, \|\tilde{W}_0\|_2\right) &\le 4 \sup_{0 \le |w| \le \pi/k} \left\{\left(1 + h + \frac{h}{k}\sin(|w|k)\right)^N \right. \\
&\qquad\qquad \left. \times \exp\left[-\frac{w^2\delta^2}{4}\right]\right\}\left(\|\tilde{F}_m\|_2 + \|\tilde{Q}_m\|_2\right) \\
&\le 4 \sup_{0 \le |w| \le \pi/k} \left\{\exp\left[hN + \frac{hN}{k}\sin(|w|k)\right]\right. \\
&\qquad\qquad \left. \times \exp\left[-\frac{w^2\delta^2}{4}\right]\right\}\left(\|\tilde{F}_m\|_2 + \|\tilde{Q}_m\|_2\right) \\
&\le 4 \sup_{0 \le |w| \le \pi/k} \left\{\exp\left[1 + |w| - \frac{w^2\delta^2}{4}\right]\right\}\left(\|\tilde{F}_m\|_2 + \|\tilde{Q}_m\|_2\right) \\
&\le 4\exp[1 + \delta^{-2}]\left(\|\tilde{F}_m\|_2 + \|\tilde{Q}_m\|_2\right),
\end{aligned}
$$

provided that $\delta \geq \sqrt{2k/\pi}$. Thus

$$\|V_0\|_2 \leq 4\exp[1 + \delta^{-2}](\|F_m\|_2 + \|Q_m\|_2) \tag{3.51}$$

and

$$\|W_0\|_2 \leq 4\exp[1 + \delta^{-2}](\|F_m\|_2 + \|Q_m\|_2) \tag{3.52}$$

by Parseval's equality. ∎

The estimates (3.51) and (3.52) show that, for a given $\delta > 0$, the finite-difference scheme (3.48)—consistent with the differential equation $J_\delta u_{xx} = (J_\delta u)_t$—is unconditionally stable and, consequently, the finite-difference solution converges to the solution of the mollified problem (3.46) by the usual arguments.

Semi-infinite Geometry

We still regard the heat equation $u_{xx} = u_t$ as a second-order differential equation when attempting to march in space. Consequently, we also need two initial conditions at $x = 1$ but, this time, we only have one, given by the approximate knowledge of the temperature $u(1, t) \approx F_m(t)$. However, the second data function can be easily obtained after solving the well-posed direct problem associated with the heat equation in the quarter plane $x > 1$, $t > 0$, with initial condition $u(x, 0) = 0$ and boundary condition $u(1, t) \approx F_m(t)—J_\delta F_m(t)$ for the mollification method. The temperature solution for the direct problem needs to be computed only at the points $(1 + h, t_j)$, $j = 1, 2, M$, in order to estimate the discrete data function $G_m(t_j) \approx u(1 + h, t_j)—J_\delta G_m(t_j)$ for the mollification method. Alternatively, it may be possible sometimes to physically implant another thermocouple at the x location $1 + h$ and directly record the temperature history values $G_m(t_j)$.

Mollification Method
There are many ways to implement a space marching finite-difference scheme when the initial data are given by two temperature history measurements. Here we describe a simple variation of the algorithm introduced in the previous section, obtained by replacing the heat fluxes W_{s-1}^j and W_s^j by their respective definitions in terms of the corresponding temperatures at the nodes.

The discrete system now reads

$$
\begin{aligned}
V_{s-1}^j &= 2V_s^j - \frac{h^2}{2k}\left(V_s^{j+1} - V_s^{j-1}\right) - V_{s+1}^j, \\
&\quad s = N, N-1, \ldots, 1, j = 1, 2, \ldots, M-N+s-1, \\
V_N^j &= J_\delta F_m(jk), &&j = 1, 2, \ldots, M, \\
V_{N+1}^j &= J_\delta G_m(jk), &&j = 1, 2, \ldots, M, \\
V_s^0 &= 0, &&s = 0, 1, \ldots, N.
\end{aligned}
\tag{3.53}
$$

The value of V_0^j, $j = 0, 1, \ldots, M - N$, is the calculated approximation for the exact temperature $f(t)$ at the grid points of the active surface $x = 0$.

Hyperbolic Method

The numerical implementation of the hyperbolic method is straightforward. Assuming that the temperature function $v(x, t)$ is sufficiently smooth, using the finite-difference quotients $(V_{s+1}^j - 2V_s^j + V_{s-1}^j)/h^2$, $(V_s^{j+1} - 2V_s^j + V_s^{j-1})/k^2$, and $(V_s^{j+1} - V_s^{j-1})/2k$ to approximate v_{xx}, v_{tt}, and v_t, respectively, the local truncation error behaves like $O(h^2 + k^2)$ as $h, k \to 0$. With h, k and $c > 1$ set to satisfy the hyperbolic stability condition $h^2 \le c^2 k^2$, substituting the finite-difference approximations in the differential equation (5.13a)—after dropping the truncation terms—we obtain the following space marching scheme:

$$V_{s-1}^j = \frac{h^2}{c^2 k^2}(V_s^{j+1} - 2V_s^j + V_s^{j-1}) + \frac{h^2}{2k}(V_s^{j+1} - V_s^{j-1}) + 2V_s^j - V_{s+1}^j,$$

$$s = N, N - 1, \ldots, 1, j = 1, 2, \ldots, M - N + s - 1,$$

$$V_N^j = F_m(jk), \qquad j = 1, 2, \ldots, M,$$

$$V_{N+1}^j = G_m(jk), \qquad j = 1, 2, \ldots, M,$$

$$V_s^0 = 0, \qquad s = 0, 1, \ldots, N + 1.$$

$$(3.54)$$

The value of V_0^j, $j = 0, 1, \ldots, M - N$, is the computed approximation for the unknown boundary temperature.

Applying the discrete Fourier transform to system (3.54), imagining the equations to hold in the entire discrete plane, we get

$$\tilde{V}_{s-1}(w) = 2\left[1 + \frac{h^2}{c^2 k^2}(\cos(wk) - 1)\right] + i\frac{h^2}{k}\sin(wk)\tilde{V}_s(w) - \tilde{V}_{s+1}(w),$$

$$\tilde{V}_N(w) = \tilde{F}_m(w),$$

$$\tilde{V}_{N+1}(w) = \tilde{G}_m(w), \qquad s = N, N - 1, \ldots, 1, 0 \le |w| \le \frac{\pi}{k},$$

or, equivalently, in matrix form,

$$\begin{pmatrix} \tilde{V}_0 \\ \tilde{V}_1 \end{pmatrix} = \begin{pmatrix} 2\left[1 + \dfrac{h^2}{c^2 k^2}(\cos(wk) - 1)\right] + i\dfrac{h^2}{k}\sin(wk) & -1 \\ 1 & 0 \end{pmatrix}^N \begin{pmatrix} \tilde{F}_m \\ \tilde{G}_m \end{pmatrix},$$

$$0 \le |w| \le \frac{\pi}{k}.$$

The discrete stability analysis—which depends on the behavior of the eigenvalues of the iteration matrix—is very similar to the one already done for the mollification method and it is left as an exercise.

Intermediate Method
We approximate the partial differential equation (5.20a) with the consistent finite-difference scheme

$$\frac{Z_{s+1}^j - 2Z_s^j + Z_{s-1}^j}{h^2}$$

$$= \frac{Z_s^{j+1} - Z_s^{j-1}}{2k} - \frac{\gamma^2\left(Z_{s+1}^{j+1} - 2Z_{s+1}^j + Z_{s+1}^{j-1} - Z_s^{j+1} + 2Z_s^j - Z_s^{j-1}\right)}{k^2 h},$$

(3.55)

$$s = N, N-1, \ldots, 1, \, j = 1, 2, \ldots, M,$$

obtaining a local truncation error that behaves as $O(h^2 + k^2)$ as $h, k \to 0$, assuming that the partial derivatives z_{xxxx}, z_{ttt}, and z_{tttt} are uniformly bounded in the quarter plane $x > 0$, $t > 0$. For computational purposes, (3.55) should be rewritten as

$$Z_{s-1}^j = \frac{h^2}{2k}\left(Z_s^{j+1} - Z_s^{j-1}\right) - \frac{\gamma^2 h}{k^2}\left(Z_{s+1}^{j+1} - 2Z_{s+1}^j + Z_{s+1}^{j-1}\right.$$

$$\left. - Z_s^{j+1} + 2Z_s^j - Z_s^{j-1}\right) - Z_{s+1}^j + 2Z_s^j,$$

$$s = N, N-1, \ldots, 1, \, j = 1, 2, \ldots, M - N + s - 1,$$

$$Z_N^j = F_m(jk), \qquad j = 1, 2, \ldots, M,$$

$$Z_{N+1}^j = G_m(jk), \qquad j = 1, 2, \ldots, M,$$

$$Z_s^0 = 0, \qquad s = 0, 1, \ldots, N+1. \tag{3.56}$$

After marching backward in space with this finite-difference scheme, the solution Z_0^j, $j = 0, 1, \ldots, M - N$, is then taken as the accepted value for the approximate boundary temperature history.

The discrete Fourier transform of system (3.56) gives, in the usual manner,

$$\begin{pmatrix} \tilde{Z}_0(w) \\ \tilde{Z}_1(w) \end{pmatrix} = \left[B_{h,k}^\gamma(w)\right]^N \begin{pmatrix} \tilde{F}_m(w) \\ \tilde{G}_m(w) \end{pmatrix}, \qquad 0 \le |w| \le \frac{\pi}{k}, \tag{3.57}$$

where

$$B^{\gamma}_{h,k}(w)$$

$$= \begin{pmatrix} 2\beta(\cos(wk) - 1) + 2(1 - i\alpha \sin(wk)) & 2\beta(1 - \cos(wk)) - 1 \\ 1 & 0 \end{pmatrix},$$

with $\alpha = h^2/2k$ and $\beta = \gamma^2 h/k^2$.

The details of the general discussion of the conditional stability of the method are left as a challenging exercise. (See Sec. 3.7 for some special cases.)

Notice that—as was the case for the hyperbolic method—we need to know two interior temperatures (at $x = 1$ and $x = 1 + h$) in order to be able to initiate the space marching scheme. Also notice that at each step—due to the structure of the computational molecule—we have to drop the estimation of the temperature from the last highest previous point in time.

3.5 INTEGRAL EQUATION APPROXIMATIONS

The simple version of the semi-infinite IHCP presented before in this chapter —system (3.1)—is equivalent to the Volterra integral equation of the first kind

$$F(t) = \int_0^t f(s) \frac{\partial \phi(1, t - s)}{\partial t} \, ds, \qquad (3.58)$$

where

$$\phi(x, t) = \mathrm{erfc}\left(\frac{x}{2\sqrt{t}} \right).$$

The corresponding Volterra integral equation for the boundary heat flux history is given by

$$F(t) = \int_0^t q(s) \frac{\partial \psi(1, t - s)}{\partial t} \, ds, \qquad (3.59)$$

where $\psi(1, t)$—the temperature response at $x = 1$ for a unit step rise of the surface heat flux at $t = 0$—is defined by

$$\psi(x, t) = \begin{cases} 2\dfrac{\sqrt{t}}{\sqrt{\pi}} \exp\left(-\dfrac{x^2}{4t} \right) - x\,\mathrm{erfc}\left(\dfrac{x}{2\sqrt{t}} \right), & t > 0, \\ 0, & t \le 0. \end{cases}$$

In the spirit of Chap. 2, (3.58) and (3.59) are both of the general abstract form (2.20) and, as such, can be solved by the regularization procedure of Tikhonov (Method IV in Chap. 2). The iterative algorithm (2.32) and the direct solution of the canonical equations (2.48) apply without changes, except that in this case the $(j - 1)$ subdiagonal of the $N \times N$ matrix A associated with the integral equation (3.45) is given by

$$a_j = \Delta\phi_j = \phi([j + 1]k) - \phi(jk), \qquad j = 1, 2, \ldots, N. \qquad (3.60)$$

These entries are obtained, for example, when using the midpoint quadrature formula to approximate (3.58) with $Nk = 1$ and $t_j = jk$, $j = 1, 2, \ldots, N$.

One of the first algorithm originally designed for the numerical solution of the IHCP—the future temperatures method (FTM)—will be discussed now. It is based on a discretization of the integral equation (3.58) and a least squares fitting of future temperature data points to effectively stabilize the ill-posed problem.

From (3.58), the exact data temperature at time t_M is approximately given by

$$F(t_M) = \sum_{n=0}^{M-1} \hat{f}_n \, \Delta\phi_{M-n} - f_M \, \Delta\phi_0, \qquad f_0 = 0, \qquad (3.61)$$

where $t_M = Mk$, $\Delta\phi_i = \phi([i + 1]k) - \phi(ik)$, $f_n = f((n - \frac{1}{2})k)$, and the caret on f_n denotes a previously estimated value of a temperature component. The only unknown in (3.61) is f_M, which is to be estimated in a sequential manner.

The temperature at time $(M + j)k$, $j = 0, 1, 2, \ldots$, is given by

$$F(t_{M+j}) = \sum_{n=0}^{M-1} \hat{f}_n \, \Delta\phi_{M-n+j} + f_M \, \Delta\phi_j + \cdots + f_{M+j} \, \Delta\phi_0. \qquad (3.62)$$

The temperature at time $(M + i - \frac{1}{2})k$, $i = 1, 2, \ldots$, may be expressed as

$$f_{M+i} = A_0 + A_1 i + A_2 i^2 + \cdots + A_\beta i^\beta \qquad (3.63)$$

and the coefficient A_0, which is equal to f_M, is to be found. For instance, if $\beta = 0$, the temperature is constant, and if $\beta = 1$, there is a linear variation of the surface temperature.

Assuming that the maximum value of j in (3.62) is equal to $r - 1$, then it is required that

$$S_r = \sum_{j=0}^{r-1} [F((M + j)k) - F_m((M + j)k)]^2 \qquad (3.64)$$

be made a minimum with respect to $A_0, A_1, \ldots, A_\beta$. To obtain a solution, r must be at least as large as $\beta + 1$ and, in order to introduce some least squares smoothing, it is necessary that $r \geq \beta + 2$. Equation (3.64) uses r future temperatures to obtain the single temperature component f_M.

Differentiation of (3.62) with respect to A_p gives the set of linear equations

$$\sum_{j=0}^{r-1} [F((M+j)k) - F((M+j)k)]C_{pj}, \qquad p = 0, 1, \ldots, \beta,$$

where

$$C_{pj} = \sum_{s=0}^{j} s^p \Delta \phi_{j-s}, \qquad p = 0, 1, \ldots, \beta.$$

This set of equations is equivalent to

$$\sum_{i=0}^{\beta} \alpha_{ip} A_i = \gamma_p, \qquad p = 0, 1, \ldots, \beta, \tag{3.65}$$

where

$$\alpha_{ip} = \alpha_{pi} = \sum_{j=0}^{r-1} c_{ij} c_{pj}, \qquad i = 0, 1, \ldots, \beta, \ p = 0, 1, \ldots, \beta,$$

$$\gamma_p = \sum_{j=0}^{r-1} F_m((M+j)k) c_{pj} - \sum_{n=0}^{M-1} \left\{ f_n \sum_{j=0}^{r-1} \Delta \phi_{M-n+j} c_{pj} \right\}, \qquad q_0 \equiv 0.$$

After k, β, and r are chosen ($r \geq \beta + 2$), the set of $\beta + 1$ linear equations given by (3.65) is solved for A_0. This solution is then taken as the accepted value for $f(t)$ over the Mth single time step only. Next, the right-hand side of (3.65) is updated, the analysis interval is shifted one time step, and the entire process is repeated. For $\beta = 0$—piecewise constant solution—there is just one linear equation with one unknown, and if $\beta = 1$—piecewise linear solution—a system of two linear equations with two unknowns is obtained.

Remarks

1. The implementation of fully explicit space marching finite-difference methods for the numerical solution of the IHCP gives maximum computational flexibility. These methods are extremely simple to program and, more importantly, they very naturally provide all the intermediate information related to the interior temperature and heat flux function values as part of the marching process itself. The same general information—using methods

based on the integral representation of the IHCP—can be obtained only by solving the direct heat conduction problem *after* the unknown boundary history temperature has been recovered. However, there are several important situations—related to free boundary problems, for example—where the location of the active surface is not known! This distinct feature of space marching schemes will play an important role in some of the applications to be studied later in Chap. 5.

2. All the methods mentioned in this chapter—with the exception of the regularization procedure of Tikhonov—are sequential, in the sense that only one component of the unknown boundary temperature and/or heat flux vector is produced at each step.

In contrast, the iterative or direct implementation of Tikhonov's method —Method IV in Chap. 2—are both, essentially, "whole domain" type of implementations. They utilize all of the data to estimate simultaneously all the components of the approximate solution vectors. Thus, if the solutions are needed over a long period of time or if the sample data frequency is very small, the dimension of the matrices and the number of computations involved increase accordingly. From this point of view, sequential methods are much more computationally efficient than whole-domain regularization methods.

3. The automatic parameter selection criterion discussed in Sec. 3.5—which determines δ in a manner consistent with the amount of noise in the data—should be used in the numerical implementation of the mollification method if a reliable upper bound, ε, for the l^2 data error is known.

4. It is always possible to combine the data filtering aspect of the mollification method with any of the other sequential methods described previously in this section. This is particularly true in the presence of very high noise, where the combined algorithms are, in general, substantially more stable and accurate when using small time sample steps. See some of the examples presented in Sec. 3.6.

3.6 NUMERICAL RESULTS

Semi-infinite Case

In order to test the actual stability and accuracy of the methods presented in the previous sections, the approximate reconstruction of a surface temperature $f(t)$ of value 1 between $t = 0.2$ and $t = 0.6$ and 0 at all other times is investigated. Such a curve has the difficult characteristics of an abrupt rise and an equally abrupt drop and constitutes a severe test because the algorithms anticipate changes in the solution temperature and also give delayed values.

We consider the data function F_m—the approximate temperature at the interior point $x = 1$—to be a discrete function measured at equally spaced sample points $t_j = jk$, $j = 0, 1, \ldots, M$, $Mk = L$ in the interval $I_L = [0, L]$, where $L = 1 + k/h$. When needed, the data functions G_m and $J_\delta G_m$—the approximate temperature at the interior point $x = 1 + h$ and the mollified approximate temperature at $x = 1 + h$—are introduced and treated in exactly the same way as $F_m(t)$ after numerically solving (just one step forward in space is needed) the corresponding well-posed direct initial value problem in the quarter plane $x > 1$, $t > 0$, with zero initial condition and boundary condition $F_m(t)$ and $J_\delta F_m(t)$, respectively.

The exact data temperature function is denoted $F(t)$ and the noisy data function $F_m(t)$ is obtained by adding random errors to $F(t_j)$. Hence, *for every grid point t_j in I_L*,

$$F_m(t_j) = F(t_j) + \varepsilon_j,$$

where the ε_j's are Gaussian random variables with variance $\sigma^2 = \varepsilon^2$.

The exact data temperature is given by

$$F(t) = \phi(t - 0.2) - \phi(t - 0.6),$$

where

$$\phi(t) = \text{erfc}\left(\frac{1}{2\sqrt{t}}\right).$$

To test the numerical stability of the algorithms, we use different average perturbations for $\varepsilon = 0.000$, 0.003, and 0.005. If the exact true component of the surface temperature solution restricted to the grid points is denoted by $f_j = f(t_j)$, we use the weighted l^2 norm—or sample root mean square norm—defined by (2.47), to measure the errors in the discretized interval $I = [0, 1]$.

Tables 3.1 and 3.2 show the results associated with the space marching finite-difference methods and the integral methods, respectively. In order, they are the mollification method (3.35) with $h/2 = k = 0.01$ and the automatic selection of δ; the hyperbolic method (3.40) with $h/2 = k = 0.01$ and quasi-optimal choices of c; the intermediate method with $h/2 = k = 0.01$ and quasi-optimal choices of γ^2; the whole-domain regularization technique of Tikhonov, implemented by solving the system of canonical equations (2.48)

TABLE 3.1. Space Marching Algorithms
(Error Norms as Functions of ε; $k = \Delta t = 0.01$, $h = \Delta x = 0.02$)

Method	Param.	$\varepsilon = 0.000$	Param.	$\varepsilon = 0.003$	Param.	$\varepsilon = 0.005$
Mollification	$\delta = k$	0.0856	$\delta = 6k$	0.1673	$\delta = 7k$	0.1903
Hyperbolic	$c = 43$	0.0855	$c = 7$	0.2053	$c = 6$	0.2230
Intermediate	$\gamma^2 = 10^{-6}$	0.0869	$\gamma^2 = 0.002$	0.1650	$\gamma^2 = 0.002$	0.1904

TABLE 3.2. Integral Equation Algorithms
(Error Norms as Functions of ε; $k = \Delta t = 0.01$)

Method	Param.	$\varepsilon = 0.000$	Param.	$\varepsilon = 0.003$	Param.	$\varepsilon = 0.005$
Canonical	$\alpha = 5 \times 10^{-5}$	0.1275	$\alpha = 6 \times 10^{-5}$	0.1302	$\alpha = 7 \times 10^{-5}$	0.1397
Iterative	$\sqrt{\alpha} = 4 \times 10^{-5}$	0.0798	$\sqrt{\alpha} = 0.004$	0.1030	$\sqrt{\alpha} = 0.005$	0.1109
Fut.-Temp.	$r = 5$	0.0698	$r = 10$	0.1230	$r = 10$	0.1364

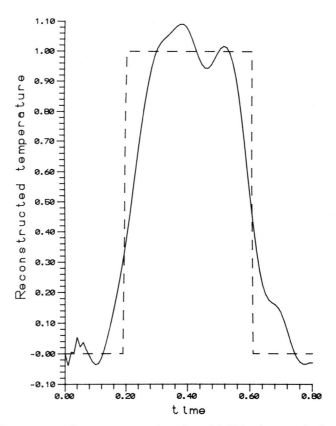

Fig. 3.1 Reconstructed temperature function. Mollification method: $\varepsilon = 0.005$, $\delta = 0.07$, $h/2 = k = 0.01$.

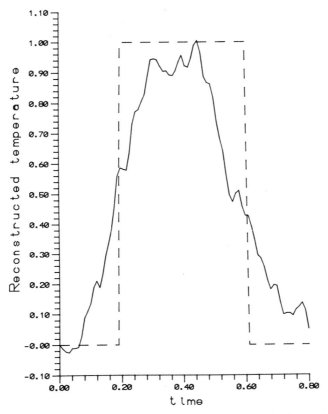

Fig. 3.2 Reconstructed temperature function. Hyperbolic method: $\varepsilon = 0.005$, $c = 6$, $h/2 = k = 0.01$.

with $N = 100$ using the midpoint quadrature formula to generate the subdiagonal entries of the matrix A [given by (3.60)]; the iterative implementation with the adjoint–conjugate gradient algorithm (2.32)—see chap. 2—with $k = 0.01$, TOL $= 1.6 \times 10^{-5}$, and initial guess $\equiv 0$; and the future temperatures method (3.65) with $k = 0.01$, constant temperature assumption ($\beta = 0$), and quasi-optimal choice of the number of future temperatures r. Convergence for the iterative scheme is achieved after 52, 22, and 22 iterations for $\varepsilon = 0.000$, 0.003, and 0.005, respectively.

Figures 3.1 to 3.6 illustrate the qualitative behavior of the reconstructed boundary temperature function with the methods mentioned previously, in that order, for $\varepsilon = 0.005$ and the other parameters as shown in the tables.

Figures 3.7 and 3.8 show the solution obtained with the combined mollification–hyperbolic method procedure with $h/2 = k = 0.01$, $\delta = 4k$, $c = 6$,

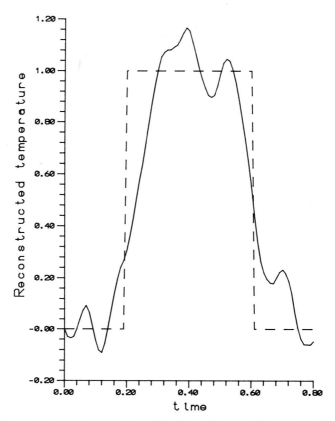

Fig. 3.3 Reconstructed temperature function. Intermediate method: $\varepsilon = 0.005$, $\gamma^2 = 0.002$, $h/2 = k = 0.01$.

and $\varepsilon = 0.005$ and the combined mollification–future temperatures method with $k = 0.01$, $\delta = 4k$, $\beta = 0$, $r = 10$, and $\varepsilon = 0.01$, respectively.

Figure 3.9 illustrates the "solution" obtained with the mollification scheme *without* filtering the initial data for $\varepsilon = 0.003$, $h/2 = k = 0.01$, and $\delta = 0$.

Remarks

1. The horizontal inspection of Tables 3.1 and 3.2 confirms the numerical stability of all the methods tested in these trials with the indicated selection of the regularization parameters. A vertical reading of the tables allows for a comparison of the accuracy of the different methods. As expected for the semi-infinite case, the integral-based methods do hold a slight edge over the space marching schemes.

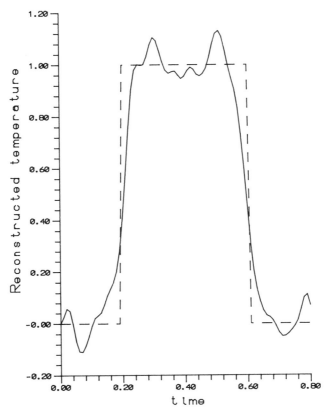

Fig. 3.4 Reconstructed temperature function. Tikhonov's method (canonical equations): $\varepsilon = 0.005$, $\alpha = 7 \times 10^{-5}$, $N = 100$.

2. The l^2 error norm of the combined mollification–hyperbolic method with $h/2 = k = 0.01$, $\delta = 4k$, $c = 6$, and $\varepsilon = 0.005$ shown in Fig. 3.7 is approximately 0.2073. This represents a somewhat modest improvement of about 10% with respect to the original value shown in Table 3.1 for the hyperbolic method. However, it is important to notice the influence that the filtering of the initial data has on the computed solution. The incipient oscillations that we observed in Fig. 3.2—indicative of some slight numerical instability—are not present at all when using the combined algorithm.

Finite Slab Case

In this section we apply the mollification method—with the finite-difference implementation given by (3.48)—to two problems in the finite slab. First, in Problem 1, the approximate reconstruction of a surface temperature $f(t)$ at

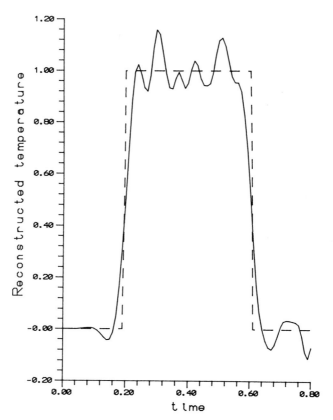

Fig. 3.5 Reconstructed temperature function. Tikhonov's method (adjoint conjugate gradient): $\varepsilon = 0.005$, $\sqrt{\alpha} = 0.005$, $N = 100$.

$x = 0$ of value 1 between $t = 0.2$ and $t = 0.6$ and of value 0 at other times is sought. The exact data for the IHCP are obtained by solving the direct problem

$$
\begin{aligned}
u_t &= u_{xx}, & 0 < x < 1, t > 0, \\
u(0, t) &= 1, & t > 0, \\
u(1, t) &= 0, & t > 0, \\
u(x, 0) &= 0, & 0 \le x \le 1.
\end{aligned}
$$

The temperature solution for this problem is given by

$$
u(x, t) = (1 - x) + \frac{2}{\pi} \sum_{n=1}^{\infty} \frac{(-1)^n}{n} \exp\left[-(n\pi)^2 t\right] \sin n\pi(1 - x), \quad t > 0,
$$

and $u(x, t) = 0$ if $t \le 0$.

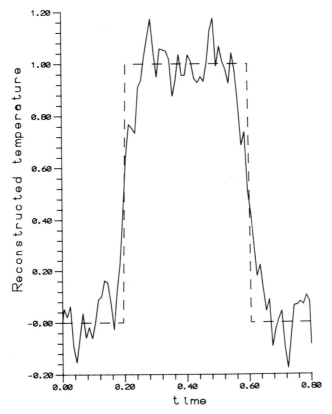

Fig. 3.6 Reconstructed temperature function. Future temperatures method: $\varepsilon = 0.005$, $r = 10$, $\beta = 0$, $k = 0.01$.

The corresponding heat flux solution is given by

$$u_x(x, t) = -\left\{ 1 + 2 \sum_{n=1}^{\infty} (-1)^n \exp\left[-(n\pi)^2 t \right] \cos n\pi(1 - x) \right\}, \qquad t > 0,$$

and $u_x(x, t) = 0$ if $t \leq 0$.

Thus, the exact data functions for the IHCP are

$$F(t) = u(1, t) = 0$$

and

$$Q(t) = u_x(1, t - 0.2) - u_x(1, t - 0.6).$$

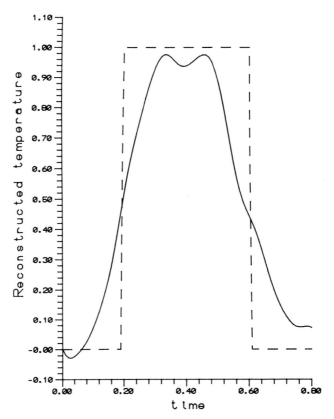

Fig. 3.7 Reconstructed temperature function. Combined mollification–hyperbolic method: $\varepsilon = 0.005$, $\delta = 4k$, $c = 6$, $h/2 = k = 0.01$.

In Problem 2 we investigate the approximate reconstruction of a surface heat flux $q(t)$ at $x = 0$ for a finite slab exposed to a nonlinear heat flux of value 0 for $t < 0.2$ with a unit jump at $t = 0.2$ and satisfying the relationship $q(t) = 1/(1 + 0.1f(t))$ for $t > 0.2$. The exact data for the IHCP are obtained by solving the direct nonlinear problem

$$(1 + 0.1u)u_t = ((1 + 0.1u)u_x)_x, \qquad 0 < x < 1, t > 0,$$

$$u_x(1, t) = 0, \qquad\qquad\qquad t > 0,$$

$$u_x(0, t) = \frac{1}{(1 + 0.1u(0, t))}, \qquad t > 0,$$

$$u(x, 0) = 0, \qquad\qquad\qquad 0 \le x \le 1.$$

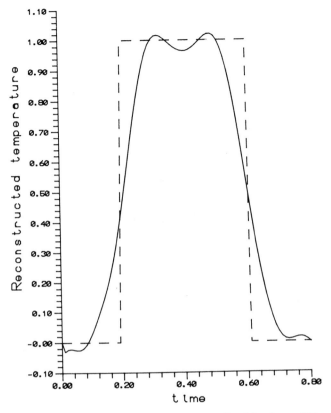

Fig. 3.8 Reconstructed temperature function. Combined mollification–future temperatures method: $\varepsilon = 0.01$, $\delta = 4k$, $r = 10$, $\beta = 0$, $k = 0.01$.

Under the transformation

$$v(x,t) = \int_0^u (1 + 0.1s)\, ds = u + \frac{1}{2}0.1u^2, \qquad (3.66)$$

the problem in the new variable v becomes

$$v_t = v_{xx}, \qquad 0 < x < 1, t > 0,$$

$$v_x(0,1) = 1, \qquad t > 0,$$

$$v_x(1,t) = 0, \qquad t > 0,$$

$$v(x,0) = 0, \qquad 0 \le x \le 1,$$

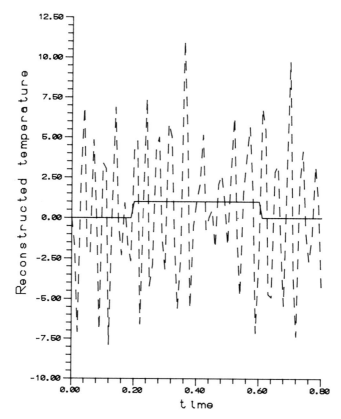

Fig. 3.9 Reconstructed temperature function. Mollification method: $\varepsilon = 0.003$, $\delta = 0$, $h/2 = k = 0.01$.

with exact temperature and heat flux solutions given respectively by

$$v(x,t) = -t - \frac{1}{6}\left[3(1-x)^2 - 1\right]$$

$$+ \frac{2}{\pi^2}\sum_{n=1}^{\infty}(-1)^n\frac{\cos n\pi(1-x)}{n^2}\exp\left[-(n\pi)^2 t\right], \qquad t > 0,$$

$$v(x,t) = 0 \qquad \text{if} \quad t \le 0, \tag{3.67}$$

and

$$v_x(x,t) = 1 - x + \frac{2}{\pi}\sum_{n=1}^{\infty}(-1)^n\frac{\sin n\pi(1-x)}{n}\exp\left[-(n\pi)^2 t\right], \qquad t > 0,$$

$$v_x(x,t) = 0 \qquad \text{if} \quad t \le 0.$$

From (3.66), we readily obtain

$$u(x,t) = \frac{1}{0.1}\left[\sqrt{1 + 0.2v(x,t)} - 1\right]. \tag{3.68}$$

In particular, the exact data functions for the nonlinear IHCP are given by

$$F(t) = u(1, t - 0.2) = 10\left(\sqrt{1 + 0.2v(1, t - 0.2)} - 1\right)$$

and

$$Q(t) = u_x(1, t - 0.2) = 0.$$

Equations (3.67) and (3.68) are also used to obtain the exact solution for the surface heat flux $q(t)$ in the IHCP which must satisfy

$$q(t) = u_x(0, t - 0.2) = \frac{1}{1 + 0.1u(0, t - 0.2)}$$

$$= \frac{1}{\sqrt{1 + 0.2v(0, t - 0.2)}}, \qquad t > 0.2,$$

and

$$q(t) = 0 \qquad \text{if} \quad t \le 0.2.$$

This function is needed for comparison purposes against our computed solution.

To numerically solve Problem 2, we rewrite the equation

$$(1 + 0.1u)u_t = \left((1 + 0.1u)u_x\right)_x$$

as

$$u_t = u_{xx} + \frac{0.1(u_x)^2}{1.0.1u}$$

and then apply the finite-difference scheme (3.48) modified as follows:

$$W_{s-1}^j = W_s^j - \frac{h}{2k}\left(V_s^{j+1} - V_s^{j-1}\right) + \frac{0.1h\left(W_s^j\right)^2}{1 + 0.1V_s^j},$$

$$V_{s-1}^j = V_s^j - hW_{s-1}^j, \qquad s = N, N-1, \ldots, 1, j = 1, 2, \ldots, M - N + s - 1,$$

$$V_s^0 = 0, \qquad\qquad s = 0, 1, \ldots, N,$$

with $h = k = 0.01$ and automatic selection of δ.

TABLE 3.3. Temperature Error Norms, Problem 1

ε	δ	Error Norm
0.000	0.01	0.0884
0.003	0.03	0.1066
0.005	0.04	0.1189

TABLE 3.4. Heat Flux Error Norms, Problem 2

ε	δ	Error Norm
0.000	0.01	0.0584
0.003	0.06	0.1208
0.005	0.07	0.1337

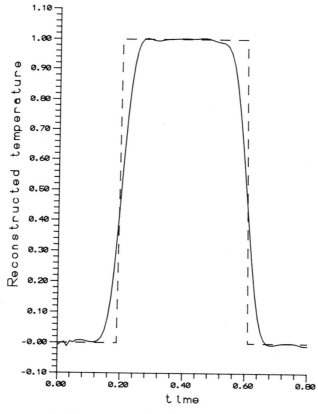

Fig. 3.10 Reconstructed temperature function. Mollification method—Problem 1: $\varepsilon = 0.005$, $\delta = 4k$, $h = k = 0.01$.

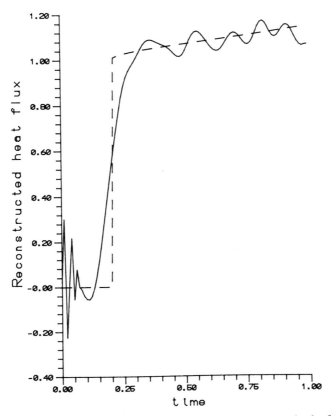

Fig. 3.11 Reconstructed heat flux function. Mollification method—Problem 2: $\varepsilon = 0.005$, $\delta = 7k$, $h = k = 0.01$.

Tables 3.3 and 3.4 show the discrete error norms—computed according to formula (2.47)—as functions of the amount of noise in the data, ε, for the computed surface temperature and surface heat flux in Problems 1 and 2, respectively.

The qualitative behavior of the reconstructed surface functions are shown in Figs. 3.10 and 3.11 for the linear and nonlinear models, respectively.

The computations do verify the theoretical stability properties of the mollification method and—in the case of the nonlinear example—show the versatility and good accuracy of the space marching explicit scheme even for small time sample intervals and a relative high level of noise in the data.

3.7 EXERCISES

3.1. Show the numerical stability of the hyperbolic method using formula (3.54) under the conditions $h^2 \le c^2 k^2$ and $h = k$.

3.2. Show that the hyperbolic system (3.54) is equivalent to the Volterra equation of the *second* kind (formally stable)

$$v(x,t) = f\left(t - \frac{x}{c}\right) \exp\left[-\frac{cx}{2}\right] + \frac{cx}{2} \int_0^{t-x/c} f(s) \exp\left[-\frac{c^2(t-s)}{2}\right]$$

$$\times \frac{1}{[(t-s)^2 - (x/c)^2]^{1/2}} I_1\left(\frac{c^2}{2}\left[(t-s)^2 - \left(\frac{x}{c}\right)^2\right]^{1/2}\right) ds, \qquad t > \frac{x}{c},$$

and

$$v(x,t) = 0, \qquad t \le \frac{x}{c},$$

where $v(0,t) = f(t)$ and

$$I_1(z) = \frac{z}{2} \sum_{k=0}^{\infty} \frac{(z/2)^{2k}}{k!(k+1)!}$$

denotes the modified Bessel function of the first kind of order 1.

3.3. Show that the intermediate method is numerically stable under the special implementation $h = k$ and $\gamma^2 = h/2$.

3.4. Replace condition (2.22) by the more general statement

$$\|Bg\|_{2,I}^2 = (Bg, Bg) = \int_0^1 [Bg(t)]^2 \, dt \le E^2.$$

(B represents, in general, a "smoothing" operator. For instance, we can have $Bg = g''$, etc.) Now choose your approximation for the function g as the one minimizing the functional

$$S_\alpha(g) = \tfrac{1}{2}\left(\|Ag - f\|_{2,I}^2 + \alpha\|Bg\|_{2,I}^2\right), \qquad \alpha > 0.$$

Find the canonical equations associated with $S_\alpha(g)$.

3.5. (*Combined future temperatures–regularization procedure*) Describe the combined algorithm obtained by requiring to minimize—with respect to

$A_0, A_1, \ldots, A_\beta$—the functional

$$S_{r,\alpha} = \sum_{j=0}^{r-1} \left[F((M+j)k) - F_m((M+j)k) \right]^2$$

$$+ \alpha \sum_{j=0}^{r-1} (q_{M+j})^2, \qquad \alpha > 0,$$

instead of (3.64).

3.6. Numerically recover the surface temperature function $f(t) = \sqrt{t}$ in the discrete time interval $I = [0, 1]$ using the mollification method with $h = k = 0.01$ and $\varepsilon = 0.001$. The exact data temperature at $x = 1$ is given by

$$F(t) = \sqrt{t} \left\{ \exp\left[-\frac{1}{4t} \right] - \frac{\sqrt{\pi}}{2\sqrt{t}} x \, \text{erfc}\left(\frac{1}{2\sqrt{t}} \right) \right\}, \qquad t > 0,$$

and

$$F(t) = 0 \qquad \text{if } t \leq 0.$$

3.7. Same as in Exercise 3.6 but solving the canonical equations from Exercise 3.4 if $Bg = g + g'$. Use $k = 0.01$ and $\varepsilon = 0.005$. Determine α by the "discrepancy" criterion (2.61).

3.8. Solve Exercise 3.6 by the combined future temperatures–regularization method of Exercise 3.5 with $h = k = 0.01$, $\beta = 1$, $r = 10$, and $\varepsilon = 0.005$. Determine α by the discrepancy principle (2.61).

3.9. Solve Exercise 3.6 by the iterative algorithm (2.32) with $k = 0.01$ and $\varepsilon = 0.005$. Use $\alpha = 10^{-2}$, 10^{-4}, and 10^{-6}.

3.8   REFERENCES AND COMMENTS

The following references and comments serve to expand the basic material covered in the corresponding sections.

In the last two decades more than 300 papers have been written about the IHCP. See Appendix B for an extensive bibliography.

3.1. An introduction and systematic overview of the IHCP with emphasis on the future temperatures method can be found in the pioneer textbook

J. V. Beck, B. F. Blackwell, and C. R. St. Clair, *Inverse Heat Conduction—Ill Posed Problems*, Wiley, New York, 1985.

Another excellent general treatise of inverse heat conduction problems and iterative techniques is available in

O. M. Alifanov, *Inverse Heat Transfer Problems*, Springer-Verlag, to appear.

For a good description of several practical aspects of the IHCP, see

E. C. Hensel, *Inverse Theory and Applications in Engineering*, Prentice-Hall, Englewood Cliffs, NJ, 1990.

3.2. Here we follow in part the work of

D. A. Murio and L. Guo, A stable marching finite difference algorithm for the inverse heat conduction problem with no initial filtering procedure, *Comput. Math. Appl.* **19** 1990, 35–50.

The mathematical concept of formal stability for the mollification method was originally introduced in

P. Manselli and K. Miller, Calculation of the surface temperature and heat flux on one side of a wall from measurements on the opposite side, *Ann. Mat. Pura Appl.* **123** (1980), 161–183.

The first extensive numerical experiments involving the integral representation of the mollified IHCP, introducing the approximated "inverse kernels" of the discrete operators, were performed by

D. A. Murio, The mollification method and the numerical solution of an inverse heat conduction problem, *SIAM J. Sci. Statist. Comput.* **2** (1981), 17–34.

For a recent generalization of the mollification method as a filtering procedure in abstract Banach spaces with stability estimates of Hölder type, the reader may consult

Dinh Nho Hào and R. Gorenflo, A non-characteristic Cauchy problem for the heat equation, *Acta Appl. Math.* **24** (1991), 1–27.

Dinh Nho Hào, A mollification method for ill-posed problems, Preprint A-91-35, Freie Universität Berlin, 1991.

The interesting suggestion of replacing the IHCP by a stable approximate hyperbolic problem is due to

C. Weber, Analysis and solution of the ill-posed inverse heat conduction problem, *Internat. J. Heat Mass Transfer* **24** (1981), 1783–1792.

A theoretical analysis of Weber's procedure can be found in

L. Eldén, Hyperbolic approximations for a Cauchy problem for the heat equation, *Inverse Problems* **4** (1988), 59–70.

For an integral solution of the IHCP based on the hyperbolic approximation method, see

D. A. Murio and C. Roth, An integral solution for the inverse heat conduction problem after the method of Weber, *Comput. Math. Appl.* **15** (1988), 39–51.

C. Roth, An integral solution approach for the inverse heat conduction problem, Ph.D. Dissertation, University of Cincinnati, 1989.

3.3–3.4. The numerical implementation of the mollification method in the finite slab case—as described in Sec. 3.4—is taken from

L. Guo, D. A. Murio, and C. Roth, A mollified space marching finite differences algorithm for the inverse heat conduction problem with slab symmetry, *Comput. Math. Appl.* **19** (1990), 75–89.

The first space marching finite-difference procedure based on a new interpretation of the mollification method by initially replacing the noisy data function F_m with the filtered data function $J_\delta F_m$ was introduced in

E. C. Hensel and R. G. Hills, A space marching finite difference algorithm for the one dimensional inverse heat conduction transfer problem, ASME Paper 84-HT-48, 1984.

The questions of numerical stability and choice of the radius of mollification for a related fully explicit space marching finite-difference method were addressed in

D. A. Murio, The mollification method and the numerical solution of the inverse heat conduction problem by finite differences, *Comput. Math. Appl.* **17** (1989), 1385–1396.

A combined space marching finite-element scheme and initial data filtering by means of Tikhonov's regularization was originally analyzed and implemented in

A. Carasso, Determining surface temperatures from interior observations, *SIAM J. Appl. Math.* **42** (1982), 558–574.

An efficient two-step mixed procedure consisting of a weighted average of a space marching step and a time marching step was presented in

M. Raynaud and J. Bransier, A new finite difference method for the nonlinear heat conduction problem, *Numer. Heat Transfer* **9** (1986), 27–42.

Several time marching finite-difference algorithms combining the numerical solution of a direct heat diffusion problem at each time step with the future temperatures method are described in the textbook by Beck, Blackwell, and St. Clair mentioned previously.

3.5. The first formulation of the future temperatures method was described in

J. V. Beck, Nonlinear estimation applied to the nonlinear inverse heat conduction problem, *Internat. J. Heat Mass Transfer* **13** (1970), 703–716.

The first application of the regularization technique of Tikhonov to the IHCP appeared in

A. N. Tikhonov and V. B. Glasko, Methods of determining the surface temperature in a body, *U.S.S.R. Computational Math. and Math. Phys.* **7** (1967), 267–273.

For a discussion of the interesting alternatives of the combined mollification–future temperatures procedure, see

D. A. Murio and J. R. Paloschi, Combined mollification–future temperatures procedure for solution of inverse heat conduction problem, *J. Comput. Appl. Math.* **23** (1988), 235–244.

An exhaustive numerical testing of the combined future temperatures–regularization procedure was presented in

J. V. Beck and D. A. Murio, Combined function specification–regularization procedure for solution of inverse heat conduction problem, *AIAA J.* **24** (1986), 180–185.

4

TWO-DIMENSIONAL INVERSE
HEAT CONDUCTION PROBLEM

Analytical and computational methods for treating the IHCP have been basically restricted to one-dimensional models. The difficulties of the two-dimensional IHCP are more pronounced and very few results are available in this case.

In this chapter we introduce a natural generalization of the mollification method for the two-dimensional inverse heat conduction problem in a slab and then concentrate our attention on the actual application of the algorithm to the more realistic situation related to a bounded rectangular domain in the (x, y) plane.

As usual, we assume all the functions involved to be L^2 functions in $\mathbb{R}^2$ suitably extended—when necessary for the analysis—as being 0 everywhere in $\{(y, t), t \leq 0, y \leq 0\}$.

4.1 TWO-DIMENSIONAL IHCP IN A SEMI-INFINITE SLAB

We consider a two-dimensional IHCP in a semi-infinite slab, in which the temperature and heat flux histories $f(y, t)$ and $q(y, t)$ on the right-hand side $(x = 1/2)$ are desired and unknown, and the temperature and heat flux on the left-hand surface $(x = 0)$ are approximately measurable.

The normalized linear problem can be described mathematically as follows. The unknown temperature $u(x, y, t)$ satisfies

$$u_t = u_{xx} + u_{yy}, \qquad 0 < x < 1/2, \, y > 0, \, t > 0, \qquad (4.1a)$$

$$u(0, y, t) = F(y, t), \qquad y > 0, \, t > 0, \qquad (4.1b)$$

with corresponding approximate data function $F_m(y, t)$,

$$u_x(0, y, t) = Q(y, t), \qquad y > 0, t > 0, \tag{4.1c}$$

with corresponding approximate data function $Q_m(y, t)$,

$$u(x, y, 0) = 0, \qquad\qquad 0 \le x \le 1/2, y \ge 0, \tag{4.1d}$$
$$u(1/2, y, t) = f(y, t), \qquad y > 0, t > 0, \tag{4.1e}$$

the desired but unknown temperature function,

$$u_x(1/2, y, t) = q(y, t), \qquad y > 0, t > 0, \tag{4.1f}$$

the desired but unknown heat flux function,

$$u(x, 0, t) = 0, \qquad 0 \le x \le 1/2, t > 0. \tag{4.1g}$$

We hypothesize that the exact data functions $F(y, t)$ and $Q(y, t)$ and the measured data functions $F_m(y, t)$ and $Q_m(y, t)$ satisfy the L^2 data error bounds

$$\|F - F_m\| \le \varepsilon \qquad \text{and} \qquad \|Q - Q_m\| \le \varepsilon. \tag{4.2}$$

Fourier transforming (4.1a) with respect to y and t, we obtain

$$\hat{u}_{xx}(x, s, w) = (s^2 + iw)\hat{u}(x, s, w), \qquad 0 < x < 1/2, -\infty < s, w < \infty. \tag{4.3}$$

Expression (4.3) is a second-order differential equation that has the general solution

$$\hat{u}(x, s, w) = A(s, w)e^{\alpha x} + B(s, w)e^{-\alpha x}, \tag{4.4}$$

where $\alpha = \sqrt{s^2 + iw}$. Differentiation of (4.4) with respect to x gives

$$\hat{u}_x(x, s, w) = A(s, w)\alpha e^{\alpha x} - B(s, w)\alpha e^{-\alpha x}. \tag{4.5}$$

Solving the system of (4.4) and (4.5) using conditions (4.1b) and (4.1c), we get

$$\begin{pmatrix} A(s, w) \\ B(s, w) \end{pmatrix} = \frac{1}{2\alpha} \begin{pmatrix} \alpha & 1 \\ \alpha & -1 \end{pmatrix} \begin{pmatrix} \hat{F}(s, w) \\ \hat{Q}(s, w) \end{pmatrix}.$$

On the other hand, according to the same equations (4.4) and (4.5), using conditions (4.1e) and (4.1f) and the preceding system, we have

$$
\begin{pmatrix} \hat{f}(s,w) \\ \\ \hat{q}(s,w) \end{pmatrix} = \begin{pmatrix} \cosh\dfrac{1}{2}\alpha & \dfrac{1}{\alpha}\sinh\dfrac{1}{2}\alpha \\ \\ \alpha\sinh\dfrac{1}{2}\alpha & \cosh\dfrac{1}{2}\alpha \end{pmatrix} \begin{pmatrix} \hat{F}(s,w) \\ \\ \hat{Q}(s,w) \end{pmatrix}. \tag{4.6}
$$

Given that

$$
\alpha = \sqrt{s^2 + iw} = \frac{1}{\sqrt{2}}\left[\left(\sqrt{s^4 + w^2} + s^2\right)^{1/2} + i\sigma\left(\sqrt{s^4 + w^2} - s^2\right)^{1/2}\right],
$$

where $\sigma = \text{sgn}(w)$, it is clear that

$$
\left|\sinh\frac{1}{2}\alpha\right| \to \infty \qquad \text{as} \quad |s| + |w| \to \infty
$$

and

$$
\left|\cosh\frac{1}{2}\alpha\right| \to \infty \qquad \text{as} \quad |s| + |w| \to \infty.
$$

Following this observation, we see that attempting to solve problem (4.1) —obtaining $f(y,t)$ and $q(y,t)$ from $F(y,t)$ and $Q(y,t)$—amplifies the error in a high-frequency component by the factor $\exp[(\sqrt{s^4 + w^2} + s^2)^{1/2}]$, showing that the inverse problem is highly ill posed in the high-frequency components.

Remark   For the one-dimensional IHCP, solving for $f(t)$ and $q(t)$ from $F(t)$ and $Q(t)$—see Sec. 3.2—the error is amplified in a high-frequency component by the factor $\exp[\sqrt{|w|/2}\,]$. Comparing this result with the previous analysis, we conclude that the two-dimensional IHCP is much more ill posed, in the high-frequency components, than the one-dimensional IHCP.

4.2   STABILIZED PROBLEM

The two-dimensional Gaussian kernel is defined by

$$
\rho(y,t,\delta_1,\delta_2) = \frac{1}{\pi\delta_1\delta_2}\exp\left[-\left(\frac{y^2}{\delta_1^2} + \frac{t^2}{\delta_2^2}\right)\right].
$$

When $\delta_1 = \delta_2 = \delta$, we denote $\rho(y,t,\delta_1,\delta_2)$ by $\rho_\delta(y,t)$. It is also true that $\rho_\delta(y,t)$ falls to nearly 0 outside the circle centered at the origin with radius

3δ and that $\int_{-\infty}^{\infty}\int_{-\infty}^{\infty}\rho_\delta(y,t)\,dy\,dt = 1$. Moreover, the Fourier transform of ρ_δ is given by

$$\hat{\rho}_\delta(s,w) = \frac{1}{2\pi}\exp\left[-\frac{\delta^2}{4}(s^2 + w^2)\right]$$

and the two-dimensional convolution of any locally integrable function $g(y,t)$ with the Gaussian kernel $\rho_\delta(y,t)$ is now written as

$$J_\delta g(y,t) = (\rho_\delta * g)(y,t) = \int_{-\infty}^{\infty}\int_{-\infty}^{\infty}\rho_\delta(y',t')g(y-y',t-t')\,dy'\,dt'.$$

Mollifying system (4.1), we obtain the following associated problem. Attempt to find $J_\delta f_m(y,t) = J_\delta u(1/2, y, t)$ and $J_\delta q_m(y,t) = J_\delta u_x(1/2, y, t)$ at some point (y,t) of interest and for some radius $\delta > 0$, given that $J_\delta u(x,y,t)$ satisfies

$$(J_\delta u)_t = (J_\delta u)_{xx} + (J_\delta u)_{yy}, \qquad 0 < x < 1/2,\ y > 0,\ t > 0,$$

$$J_\delta u(0, y, t) = J_\delta F_m(y, t), \qquad y > 0,\ t > 0,$$

$$J_\delta u_x(0, y, t) = J_\delta Q_m(y, t), \qquad y > 0,\ t > 0, \qquad (4.7)$$

$$J_\delta u(x, y, 0) = 0, \qquad 0 \le x \le 1/2,\ y \ge 0,$$

$$J_\delta u(x, 0, t) = 0, \qquad 0 \le x \le 1/2,\ t > 0.$$

This problem and its solutions satisfy the following theorem.

Theorem 4.1   Suppose that $\|F - F_m\| \le \varepsilon$ and $\|Q - Q_m\| \le \varepsilon$. Then

(i) Problem (4.7) is a formally stable problem with respect to perturbations in the data.

(ii) If the exact boundary temperature $f(y,t)$ and heat flux $q(y,t)$ have uniformly bounded first-order partial derivatives on the bounded domain $D = [0, Y] \times [0, T]$, then $J_\delta f_m$ and $J_\delta q_m$ verify

$$\|f - J_\delta f_m\|_D \le O(\delta) + 2\varepsilon \exp[7\delta^{-2}]$$

and

$$\|q - J_\delta q_m\|_D \le O(\delta) + 2\varepsilon \exp[7\delta^{-2}].$$

Proof

(i) Repeating the same procedure as the one used to obtain (4.6), we have

$$
\begin{pmatrix} \left(\widehat{J_\delta f}\right) - \left(\widehat{J_\delta f_m}\right) \\[2mm] \left(\widehat{J_\delta q}\right) - \left(\widehat{J_\delta q_m}\right) \end{pmatrix}
$$

$$
= \begin{pmatrix} \cosh\dfrac{1}{2}\alpha & \dfrac{1}{\alpha}\sinh\dfrac{1}{2}\alpha \\[3mm] \alpha \sinh\dfrac{1}{2}\alpha & \cosh\dfrac{1}{2}\alpha \end{pmatrix} \exp\left[-\dfrac{s^2 + w^2}{4}\delta^2\right] \begin{pmatrix} \hat{F} - \hat{F}_m \\[2mm] \hat{Q} - \hat{Q}_m \end{pmatrix}.
$$

In particular,

$$
\left\|\left(\widehat{J_\delta f}\right) - \left(\widehat{J_\delta f_m}\right)\right\| \le \max_{|w|,\,|s|} \left|\cosh\dfrac{1}{2}\alpha \exp\left[-\dfrac{s^2 + w^2}{4}\delta^2\right]\right| \|\hat{F} - \hat{F}_m\|
$$

$$
\tag{4.8}
$$

$$
+ \max_{|w|,\,|s|} \left|\dfrac{1}{\alpha}\sinh\dfrac{1}{2}\alpha \exp\left[-\dfrac{s^2 + w^2}{4}\delta^2\right]\right| \|\hat{Q} - \hat{Q}_m\|
$$

and

$$
\left\|\left(\widehat{J_\delta q}\right) - \left(\widehat{J_\delta q_m}\right)\right\| \le \max_{|w|,\,|s|} \left|\alpha \sinh\dfrac{1}{2}\alpha \exp\left[-\dfrac{s^2 + w^2}{4}\delta^2\right]\right| \|\hat{F} - \hat{F}_m\|
$$

$$
+ \max_{|w|,\,|s|} \left|\cosh\dfrac{1}{2}\alpha \exp\left[-\dfrac{s^2 + w^2}{4}\delta^2\right]\right| \|\hat{Q} - \hat{Q}_m\|. \tag{4.9}
$$

We notice that

$$
\left|\cosh\dfrac{1}{2}\alpha\right| \le e^{|\alpha|} \le \exp\left(|s| + \sqrt{|w|}\,\right),
$$

$$
\left|\dfrac{1}{\alpha}\sinh\dfrac{1}{2}\alpha\right| \le e^{|\alpha|} \le \exp\left(|s| + \sqrt{|w|}\,\right),
$$

and

$$
\left|\alpha \cosh\dfrac{1}{2}\alpha\right| \le |\alpha|e^{|\alpha|} \le \exp\left(2|s| + 2\sqrt{|w|}\,\right).
$$

Thus,

$$
\max_{|w|,|s|} \left| \alpha \sinh \frac{1}{2}\alpha \exp\left[-\frac{s^2+w^2}{4}\delta^2\right] \right|
$$

$$
\leq \max_{|w|,|s|} \left\{ \exp\left[2|s| - \frac{s^2\delta^2}{4}\right] \exp\left[-\frac{s^2+w^2}{4}\delta^2\right] \right\}
$$

$$
\leq \max_{|s|} \exp\left[2|s| - \frac{s^2\delta^2}{4}\right] \max_{|w|} \exp\left[2|w| - \frac{\sqrt{2}\,\delta^2}{4}\right]
$$

$$
\leq \exp(4\delta^{-2})\exp(3\delta^{-2/3}).
$$

This is a uniform upper bound for all the related factors in expressions (4.8) and (4.9). Consequently,

$$
\left\| \widehat{(J_\delta f)} - \widehat{(J_\delta f_m)} \right\| \leq \exp(4\delta^{-2})\exp(3\delta^{-2/3})\left(\|\hat{F} - \hat{F}_m\| + \|\hat{Q} - \hat{Q}_m\| \right)
$$

and

$$
\left\| \widehat{(J_\delta q)} - \widehat{(J_\delta q_m)} \right\| \leq \exp(4\delta^{-2})\exp(3\delta^{-2/3})\left(\|\hat{F} - \hat{F}_m\| + \|\hat{Q} - \hat{Q}_m\| \right).
$$

Using Parseval's equality and the error bound for the data, we obtain

$$
\|J_\delta f - J_\delta f_m\|_D \leq \|J_\delta f - J_\delta f_m\| \leq 2\varepsilon \exp(4\delta^{-2})\exp(3\delta^{-2/3})
$$

$$
\leq 2\varepsilon \exp(7\delta^{-2}) \tag{4.10}
$$

and

$$
\|J_\delta q - J_\delta q_m\|_D \leq \|J_\delta q - J_\delta q_m\| \leq 2\varepsilon \exp(4\delta^{-2})\exp(3\delta^{-2/3})
$$

$$
\leq 2\varepsilon \exp(7\delta^{-2}). \tag{4.11}
$$

This shows that the mollified problem (4.7) is a formally stable problem with respect to perturbations in the data. By the preceding inequalities, for a fixed $\delta > 0$, the error is guaranteed to tend to 0 as $\varepsilon \to 0$.

(ii) Consider

$$
f(y,t) - J_\delta f_m(y,t) = f(y,t) - J_\delta f(y,t) + J_\delta f(y,t) - J_\delta f_m(y,t)
$$

and

$$
q(y,t) - J_\delta q_m(y,t) = q(y,t) - J_\delta q(y,t) + J_\delta q(y,t) - J_\delta q_m(y,t).
$$

Since, by assumption, all the partial derivatives of f and q are uniformly bounded in D, both differences $\|f - J_\delta f\|_D$ and $\|q - J_\delta q\|_D$ are bounded by a constant times δ [see Sec. 3.2, (3.10)]. Combining this with the estimates (4.10) and (4.11) and using the triangle inequality, we immediately have

$$\|f - J_\delta f_m\|_D \leq O(\delta) + 2\varepsilon \exp(7\delta^{-2})$$

and

$$\|q - J_\delta q_m\|_D \leq O(\delta) + 2\varepsilon \exp(7\delta^{-2}.) \quad \blacksquare$$

4.3   NUMERICAL PROCEDURE AND ERROR ANALYSIS

With $v = J_\delta u$ and $w = \partial v / \partial x$, system (4.7) is equivalent to

$$\frac{\partial v}{\partial t} = \frac{\partial w}{\partial x} + \frac{\partial^2 v}{\partial y^2}, \qquad 0 < x < 1/2, \, y > 0, \, t > 0,$$

$$w = \frac{\partial v}{\partial x}, \qquad 0 < x < 1/2, \, y > 0, \, t > 0,$$

$$v(0, y, t) = J_\delta F_m(y, t), \qquad y > 0, \, t > 0,$$

$$w(0, y, t) = J_\delta Q_m(y, t), \qquad y > 0, \, t > 0, \qquad (4.12)$$

$$v(x, y, 0) = 0, \qquad 0 \leq x \leq 1/2, \, y \geq 0,$$

$$v(1/2, y, t) = J_\delta f_m(y, t), \qquad y > 0, \, t > 0, \text{ unknown,}$$

$$w(1/2, y, t) = J_\delta q_m(y, t), \qquad y > 0, \, t > 0, \text{ unknown,}$$

$$v(x, 0, t) = 0, \qquad 0 \leq x \leq 1/2, \, t > 0.$$

Without loss of generality, we will seek to reconstruct the unknown mollified boundary temperature function $J_\delta f_m$ or mollified boundary heat flux function $J_\delta q_m$ in the unit square $D = [0, 1] \times [0, 1]$ of the (y, t) plane $x = 1/2$. Consider a uniform grid in the (x, y, t) space:

$$\{(x_i = ih, \, y_j = js, \, t_n = nk), \, i = 0, 1, \ldots, N, \, Nh = 1/2;$$

$$j = 0, 1, \ldots, M, \, Ms = L; \, n = 0, 1, \ldots, P, \, Pk = C\},$$

where L and C depend on h, s, and k in a way to be specified later, $L, C > 1$. Let the grid functions V and W be defined by

$$V_{i,j}^n = v(x_i, y_j, t_n), \qquad W_{i,j}^n = w(x_i, y_j, t_n),$$

$$0 \leq i \leq N, 0 \leq j \leq M, 0 \leq n \leq P.$$

We notice that

$$V_{0,j}^n = J_\delta F_m(y_j, t_n), \qquad 0 \le j \le M, 0 \le n \le P,$$

$$W_{0,j}^n = J_\delta Q_m(y_j, t_n), \qquad 0 \le j \le M, 0 \le n \le P,$$

$$V_{i,0}^n = 0, \qquad 0 \le i \le N, 0 \le n \le P,$$

$$V_{i,j}^0 = 0, \qquad 0 \le i \le N, 0 \le j \le M.$$

We approximate the system of partial differential equations (4.12) with the consistent finite-difference schemes

$$W_{i+1,j}^n = W_{i,j}^n + \frac{h}{2k}\left(V_{i,j}^{n+1} - V_{i,j}^{n-1}\right) - \frac{h}{s^2}\left(V_{i,j-1}^n - 2V_{i,j}^n + V_{i,j+1}^n\right),$$

$$V_{i+1,j}^n = V_{i,j}^n + hW_{i+1,j}^n,$$

$$i = 0,1,\ldots,N-1,\, j = 1,2,\ldots,M-i-1,\, n = 1,2,\ldots,P-i-1,$$

$$V_{0,j}^n = (J_\delta F_m)_j^n, \qquad j = 1,2,\ldots,M-1,\, n = 1,2,\ldots,P-1,$$

$$W_{0,j}^n = (J_\delta Q_m)_j^n, \qquad j = 1,2,\ldots,M-1,\, n = 1,2,\ldots,P-1,$$

$$V_{i,0}^n = 0, \qquad i = 0,1,\ldots,N-1,\, n = 1,2,\ldots,P-i, \qquad (4.13)$$

$$V_{i,j}^0 = 0, \qquad i = 1,2,\ldots,N,\, j = 1,2,\ldots,M.$$

Notice that as we march forward in the x direction in space, we must drop the estimation of the interior temperature from the highest previous point in time and the associated rightmost point in the y direction. Since we want to evaluate $\{V_{N,j}^n\}$ and $\{W_{N,j}^n\}$ at the grid points of the unit square $D = [0,1] \times [0,1]$ after N iterations, the minimum initial length C of the data sample interval in the time axis needs to satisfy the condition $C = Pm = 1 - k + k/h$. Similarly, the minimum initial length L of the data sample interval in the y direction satisfies $L = Ms = 1 - s + s/h$.

Now, in order to analyze the stability of the finite-difference scheme (4.13) and the convergence of the numerical solution to the solution of the mollified problem (4.7), we use the discrete Fourier transform method once again. The definition of a two-dimensional discrete Fourier transform of a grid function $\{U_j^n\}_{j,n=1}^\infty$—similar to the one-dimensional discrete Fourier transform introduced in Definition A.16—is given by

$$\tilde{U}(w_1, w_2) = \frac{sk}{2\pi} \sum_{n=-\infty}^{\infty} \sum_{j=-\infty}^{\infty} U_j^n e^{i(jw_1 s + nw_2 k)},$$

$$0 \le |w_1| \le \frac{\pi}{s}, \qquad 0 \le |w_2| \le \frac{\pi}{k}, \qquad i = \sqrt{-1}.$$

Applying this transformation to (4.13), we get

$$\tilde{W}_{i+1} = \tilde{W}_i + \frac{h}{2k}\left(\tilde{V}_i e^{iw_2 k} - \tilde{V}_i e^{-iw_2 k}\right) - \frac{h}{s^2}\left(\tilde{V}_i e^{iw_1 s} - 2\tilde{V}_i + \tilde{V}_i e^{-iw_1 s}\right),$$

$$\tilde{V}_{i+1} = \tilde{V}_i + h\tilde{W}_{i+1}.$$

Equivalently, we have

$$\begin{pmatrix} \tilde{V}_{i+1} \\ \tilde{W}_{i+1} \end{pmatrix} = \begin{pmatrix} 1 + ha & h \\ a & 1 \end{pmatrix}\begin{pmatrix} \tilde{V}_i \\ \tilde{W}_i \end{pmatrix},$$

where

$$a = \frac{4h}{s^2}\left(\sin\frac{w_1 s}{2}\right)^2 + i\frac{h}{k}\sin w_2 k.$$

Thus,

$$|\tilde{V}_{i+1}| \le \{1 + h + h|a|\}\max\left(|\tilde{V}_i|, |\tilde{W}_i|\right)$$

and

$$|\tilde{W}_{i+1}| \le \{1 + |a|\}\max\left(|\tilde{V}_i|, |\tilde{W}_i|\right).$$

Consequently,

$$\max\left(|\tilde{V}_{i+1}|, |\tilde{W}_{i+1}|\right) \le \{1 + |a|\}\max\left(|\tilde{V}_i|, |\tilde{W}_i|\right)$$

and

$$\begin{aligned}
\max\left(|\tilde{V}_N|, |\tilde{W}_N|\right) &\le \{1 + |a|\}^N \max\left(|\tilde{V}_0|, |\tilde{W}_0|\right) \\
&= \{1 + |a|\}^N \max\left(|\tilde{\rho}_\delta \tilde{V}_0|, |\tilde{\rho}_\delta \tilde{W}_0|\right) \\
&= \{1 + |a|\}^N |\tilde{\rho}_\delta| \max\left(|\tilde{F}_m|, |\tilde{Q}_m|\right).
\end{aligned}$$

From the last inequality it follows that

$$|\tilde{V}_N|^2 \le \{1 + |a|\}^{2N}|\tilde{\rho}_\delta|^2\left(|\tilde{F}_m|^2 + |\tilde{Q}_m|^2\right) \tag{4.14a}$$

and

$$|\tilde{W}_N|^2 \le \{1 + |a|\}^{2N}|\tilde{\rho}_\delta|^2\left(|\tilde{F}_m|^2 + |\tilde{Q}_m|^2\right). \tag{4.14b}$$

A similar analysis to the one done in Sec. 3.3, using Poisson's summation formula in (4.14a) and (4.14b), integrating with respect to w_1 and w_2, and taking square roots, gives

$$
\max\left(\|\tilde{V}_N\|_2, \|\tilde{W}_N\|_2\right) \leq 4 \sup_{\substack{0<|w_1|\leq \pi/k \\ 0<|w_2|\leq \pi/s}} \left\{ \{1 + |a|\}^N \right.
$$

$$
\left. \times \exp\left[-\frac{(w_1^2 + w_2^2)\delta^2}{4}\right]\right\} \left(\|\tilde{F}_m\|_2 + \|\tilde{Q}_m\|_2\right)
$$

$$
\leq 4 \sup_{\substack{0<|w_1|\leq \pi/k \\ 0<|w_2|\leq \pi/s}} \left\{ \left\{1 + \frac{4h}{s^2}\frac{\left(|w_1|^{1/2}s\right)^2}{4} + \frac{h}{k}|w_2|k\right\}^N \right.
$$

$$
\left. \times \exp\left[-\frac{(w_1^2 + w_2^2)\delta^2}{4}\right]\right\} \left(\|\tilde{F}_m\|_2 + \|\tilde{Q}_m\|_2\right)
$$

$$
\leq 4 \sup_{\substack{0<|w_1|\leq \pi/k \\ 0<|w_2|\leq \pi/s}} \left\{ \{1 + h(|w_1| + |w_2|)\}^N \right.
$$

$$
\left. \times \exp\left[-\frac{(w_1^2 + w_2^2)\delta^2}{4}\right]\right\} \left(\|\tilde{F}_m\|_2 + \|\tilde{Q}_m\|_2\right)
$$

$$
\leq 4 \sup_{\substack{0<|w_1|\leq \pi/k \\ 0<|w_2|\leq \pi/s}} \left\{ \exp\left[Nh(|w_1| + |w_2|)\right] \right.
$$

$$
\left. \times \exp\left[-\frac{(w_1^2 + w_2^2)\delta^2}{4}\right]\right\} \left(\|\tilde{F}_m\|_2 + \|\tilde{Q}_m\|_2\right)
$$

$$
\leq 4 \sup_{0<|w_1|\leq \pi/k} \left\{ \exp\left[\frac{1}{2}|w_1| - w_1^2\frac{\delta^2}{4}\right]\right\}
$$

$$
\times 4 \sup_{0<|w_2|\leq \pi/s} \left\{ \exp\left[\frac{1}{2}|w_2| - w_2^2\frac{\delta^2}{4}\right]\right\}
$$

$$
\times \left(\|\tilde{F}_m\|_2 + \|\tilde{Q}_m\|_2\right)
$$

$$
\leq 16 \exp\left[\frac{1}{4\delta^2}\right] \left(\|\tilde{F}_m\|_2 + \|\tilde{Q}_m\|_2\right),
$$

provided that $\delta \geq \max(\sqrt{k/\pi}, \sqrt{s/\pi})$.

Thus,

$$\|V_N\|_2 \le 16 \exp\left[\frac{1}{4\delta^2}\right](\|F_m\|_2 + \|Q_m\|_2) \tag{4.15}$$

and

$$\|W_N\|_2 \le 16 \exp\left[\frac{1}{4\delta^2}\right](\|F_m\|_2 + \|Q_m\|_2) \tag{4.16}$$

by Parseval's equality. ∎

The estimates (4.15) and (4.16) show that for a given $\delta > 0$, the finite-difference scheme (4.13)—consistent with the partial differential equation $J_\delta u_{xx} = (J_\delta u)_t - J_\delta u_{yy}$—is unconditionally stable and the finite-difference solution converges to the solution of the mollified problem (4.7).

Remarks

1. The radius of mollification, δ, can be selected automatically as a function of the level of noise in the data. The selection criterion discussed in Sec. 3.5 extends naturally to the two-dimensional case. In fact, for a given $\varepsilon > 0$, there is a unique $\delta > 0$, such that

$$\|J_\delta F_m - F_m\|_D = \varepsilon. \tag{4.17}$$

2. From a more practical point of view, it is possible to replace the discrete two-dimensional mollification of the data functions by two successive one-dimensional mollifications in time and y space. In this manner, the data filtering task can be executed as a parallel process, eliminating the most time-consuming aspect of the algorithm.

4.4 NUMERICAL RESULTS

Two-Dimensional Strip

We test the accuracy and the stability properties of the method on two model problems.

Problem 1   The approximate reconstruction of a surface temperature $f(y,t)$ and a surface heat flux $q(y,t)$ at $x = 1/2$ are investigated for a two-dimensional strip exposed to a heat flux data function at the free surface $x = 0$, given by

$$u_x(0, y, t) = Q(y, t)$$

$$= \begin{cases} \left[1 + 2 \sum_{n=1}^{\infty} (-1)^n \exp(-n^2\pi^2 t)\right] \mathrm{erfc}\left(\dfrac{y}{2\sqrt{t}}\right), & y \geq 0, t > 0, \\ 0, & \text{otherwise}, \end{cases}$$

and a temperature data function

$$u(0, y, t) = F(y, t) = 0 \qquad \text{for all } t \text{ and } y.$$

The exact temperature solution to be approximately reconstructed at the surface $x = 1/2$ has equation

$$u\left(\frac{1}{2}, y, t\right) = f(y, t)$$

$$= \begin{cases} \left\{\dfrac{1}{2} + \dfrac{2}{\pi} \sum_{n=1}^{\infty} \dfrac{(-1)^n}{2n-1} \exp[(-2n-1)^2\pi^2 t]\right\} \mathrm{erfc}\left(\dfrac{y}{2\sqrt{t}}\right), & y \geq 0, t > 0, \\ 0, & \text{otherwise}, \end{cases}$$

and the exact heat flux solution at the surface $x = 1/2$ is given by

$$u_x\left(\frac{1}{2}, y, t\right) = q(y, t)$$

$$= \begin{cases} \left\{1 + 2 \sum_{n=1}^{\infty} (-1)^n \exp[-(2n)^2\pi^2 t]\right\} \mathrm{erfc}\left(\dfrac{y}{2\sqrt{t}}\right), & y \geq 0, t > 0, \\ 0, & \text{otherwise}. \end{cases}$$

Problem 2   We attempt to approximately reconstruct a surface temperature $f(y, t)$ and a surface heat flux $q(y, t)$ at $x = 1/2$ for a two-dimensional slab subject to a nonlinear diffusion process. The IHCP is described as follows:

$$u_t = \frac{x-1}{t} \frac{u}{u_x} u_{xx} + u_{yy}, \qquad 0 < x < 1/2, y > 0, 0 < t < T,$$

$$u_x(0, y, t) = -2te^t e^{-y} = Q(y, t), \qquad y > 0, 0 < t < T,$$

$$u(0, y, t) = te^t e^{-y} = F(y, t), \qquad y > 0, 0 < t < T,$$

$$u(x, y, 0) = 0, \qquad 0 \leq x \leq 1/2, y \geq 0, \qquad (4.18)$$

$$u(1/2, y, t) = f(y, t), \quad \text{unknown}, \qquad y > 0, 0 < t < T,$$

$$u_x(1/2, y, t) = q(y, t), \quad \text{unknown}, \qquad y > 0, 0 < t < T,$$

$$u(x, 0, t) = (1-x)^2 te^t = h(x, t), \qquad 0 \leq x \leq 1/2, 0 < t < T,$$

$$u(x, y, t) \to 0 \quad \text{as} \quad y \to \infty, \qquad 0 < x < 1/2, 0 < t < T.$$

The unique temperature solution of the associated direct problem is

$$u(x, y, t) = (1 - x)^2 te^t e^{-y}.$$

Hence, the exact solutions for the nonlinear IHCP are given by

$$f(y, t) = \tfrac{1}{4} te^t e^{-y}, \qquad y \geq 0, 0 \leq t \leq T,$$

and

$$q(y, t) = -te^t e^{-y}, \qquad y \geq 0, 0 \leq t \leq T.$$

In order to approximate the nonlinear partial differential equation in (4.18), we rewrite the first equation of the finite-difference scheme (4.13) as

$$W_{i+1, j}^n = W_{i, j}^n + \frac{t_j}{x_i - 1} \frac{W_{i, j}^n}{V_{i, j}^n} \left[\frac{h}{2k} \left(V_{i, j}^{n+1} - V_{i, j}^{n-1} \right) \right.$$

$$\left. - \frac{h}{s^2} \left(V_{i, j-1}^n - 2V_{i, j}^n + V_{i, j+1}^n \right) \right].$$

We assume the data functions F_m and Q_m to be discrete functions measured at equally spaced points in the (y, t) domain $[0, L] \times [0, C]$, where $L = 1 - s + s/h$, $C = 1 - k + k/h$, $Nh = 1/2$, $h = \Delta x$, $s = \Delta y$, and $k = \Delta t$. The $(M + 1) \times (P + 1)$ sample points in this rectangle have coordinates $(y_j, t_n) = (js, nk)$, $0 \leq j \leq M$, $0 \leq n \leq P$, $Ms = L$, $Pn = C$. After extending the data functions in such a way that they decay smoothly to 0 outside $[0, L] \times [0, C]$, we consider the data functions F_m and Q_m defined at equally spaced sample points on any rectangular domain of interest in the (y, t) plane.

Once the radii of mollification δ_F and δ_Q, associated with the data functions F_m and Q_m, respectively, have been obtained—after solving the discrete version of (4.17)—and the discrete filtered data functions

$$J_{\delta_F} F_m(y_j, t_n) = V_{0, j}^n, \qquad J_{\delta_Q} Q_m(y_j, t_n) = W_{0, j}^n, \qquad 0 \leq j \leq M, 0 \leq n \leq P,$$

are determined, we apply the finite-difference algorithm described in Sec. 4.3, marching forward in the x direction. The values $V_{N, j}^n$ and $W_{N, j}^n$, $0 \leq j \leq M - N$, $0 \leq n \leq P - N$, so obtained, are then taken as the accepted approximations for the boundary temperature and heat flux histories respectively at the different y locations in the domain $D = [0, 1] \times [0, 1]$ of the (y, t) plane at $x = 1/2$.

In Problem 1 we use $h = s = 0.1$ and $k = 0.01$. In Problem 2 we choose $h = 0.1$, $s = 0.015$, and $k = 0.01$. If the exact data temperature (heat flux) is denoted by $F(y, t) [Q(y, t)]$, the noisy data $F_m(y_j, t_n) [Q_m(y_j, t_n)]$ is obtained by adding a random error to F_m [Q_m]; that is, for every grid point (y_j, t_n),

$$F_m(y_j, t_n) = F(y_j, t_n) + \varepsilon_{1,j,n},$$

$$Q_m(y_j, t_n) = Q(y_j, t_n) + \varepsilon_{2,j,n},$$

where $\varepsilon_{1,j,n}$ and $\varepsilon_{2,j,n}$ are Gaussian random variables of variance $\varepsilon^2 = (\varepsilon_1)^2 = (\varepsilon_2)^2$.

To study the numerical stability of the algorithm, we use different average perturbations for $\varepsilon = 0$, 0.001, 0.002, 0.003, 0.004, and 0.005. The solution errors for the discrete temperature and heat flux functions at $x = 1/2$ in the discretized time interval $I = [0, 1]$, at the $y = 1/2$ location, are respectively given by

$$\|V_{N,j} - f_j\|_I = \left\{ \frac{1}{P - N + 1} \sum_{n=1}^{P-N+1} (V_{N,j}^n - f_j^n)^2 \right\}^{1/2}$$

and

$$\|Q_{N,j} - q_j\|_I = \left\{ \frac{1}{P - N + 1} \sum_{n=1}^{P-N+1} (W_{N,j}^n - q_j^n)^2 \right\}^{1/2}, \qquad js = \frac{1}{2}.$$

Tables 4.1 and 4.2 show the results of the numerical experiments associated with Problems 1 and 2, respectively. The qualitative behavior of the reconstructed surface temperature for Problem 1 (at y location $1/2$) is illustrated in Fig. 4.1a where the numerical solution for $\varepsilon = 0.005$ is plotted. The corresponding plot for the heat flux function is shown in Fig. 4.1b. Figure 4.2a shows the reconstructed surface temperature (y location $1/2$) for Problem 2 corresponding to the noise level $\varepsilon = 0.005$. The associated heat flux is plotted in Fig. 4.2b.

TABLE 4.1. Error Norm as a Function of the Amount of Noise in the Data for the Surface Temperature and Surface Heat Flux in Problem 1 at $x = y = 1/2$

Temperature			Heat Flux		
ε	δ_F	Error Norm	ε	δ_Q	Error Norm
0.000	0.01	0.003580	0.000	0.01	0.017352
0.001	0.04	0.003580	0.001	0.04	0.017356
0.002	0.04	0.003581	0.002	0.04	0.017364
0.003	0.04	0.003583	0.003	0.04	0.017374
0.004	0.04	0.003585	0.004	0.04	0.017388
0.005	0.04	0.003588	0.005	0.04	0.017404

TABLE 4.2. Error Norm as a Function of the Amount of Noise in the Data for the Surface Temperature and Surface Heat Flux in Problem 2 at $x = y = 1/2$

Temperature			Heat Flux		
ε	δ_F	Error Norm	ε	δ_Q	Error Norm
0.000	0.01	0.010779	0.000	0.01	0.043116
0.001	0.04	0.010831	0.001	0.04	0.043327
0.002	0.04	0.011099	0.002	0.04	0.043462
0.003	0.04	0.011341	0.003	0.04	0.044398
0.004	0.04	0.011343	0.004	0.04	0.044428
0.005	0.04	0.011345	0.005	0.04	0.045371

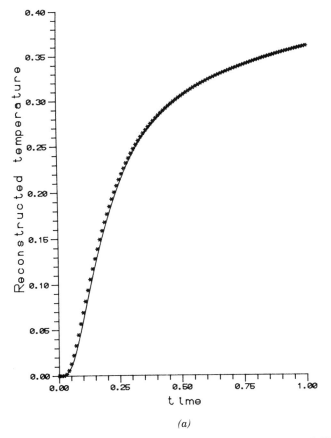

(a)

Fig. 4.1a   Reconstructed temperature function—Problem 1: $x = y = 1/2$, $\varepsilon = 0.005$, $\delta = 0.04$, $h = s = 0.1$, $k = 0.01$. Exact solution ($\ast\ast\ast$); computed solution (—).

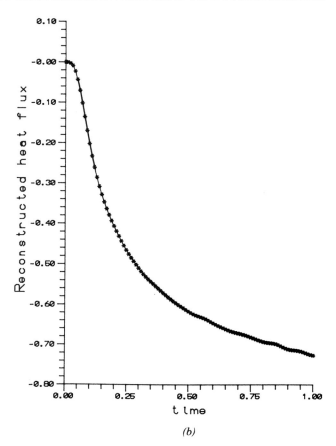

(b)

Fig. 4.1b   Reconstructed heat flux function—Problem 1: $x = y = 1/2$, $\varepsilon = 0.005$, $\delta = 0.04$, $h = s = 0.1$, $k = 0.01$. Exact solution ($* * *$); computed solution (—).

Bounded Rectangular Domain

We now investigate the two-dimensional IHCP when the space domain in the (x, y) plane is restricted to the bounded prototype rectangle $R = [0, 1/2] \times [0, 1]$. We must add the "boundary" condition—from the point of view of the IHCP and the x-space marching scheme—

$$u(x, 1, t) = h_1(x, t), \qquad 0 \le x \le 1/2, t > 0,$$

and, at the same time, we shall not require the homogeneity of the boundary condition (4.1g) that should now read

$$u(x, 0, t) = h_0(x, t), \qquad 0 \le x \le 1/2, t > 0.$$

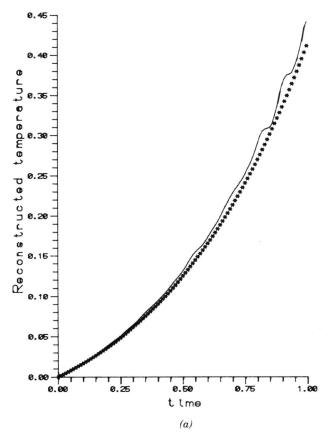

(a)

Fig. 4.2a Reconstructed temperature function—Problem 2: $x = y = 1/2$, $\varepsilon = 0.005$, $\delta = 0.04$, $h = 0.1$, $s = 0.015$, $k = 0.01$. Exact values ($\ast \ast \ast$); computed solution (—).

However, since the functions f, h_0, h_1, and F uniquely determine the solution of the direct problem—including the heat fluxes Q and q at $x = 0$ and $x = 1/2$, respectively—it is clear that, for the inverse problem, the data functions F and Q possess all the necessary information for the recovery of the temperature and heat flux functions f and q at the surface $x = 0$ and also the boundary conditions h_0 and h_1 at $y = 0$ and $y = 1$, respectively. Consequently, the functions $u(x, 0, t) = h_0(x, t)$ and $u(x, 1, t) = h_1(x, t)$ will be treated as *unknowns* and their recovery becomes a natural task for the two-dimensional IHCP. Mathematically, the new inverse problem can be stated as follows:

$$u_t = u_{xx} + u_{yy}, \qquad 0 < x < 1/2, 0 < y < 1, t > 0, \quad (4.19a)$$

$$u(0, y, t) = F(y, t), \qquad 0 < y < 1, t > 0, \qquad\qquad\qquad (4.19b)$$

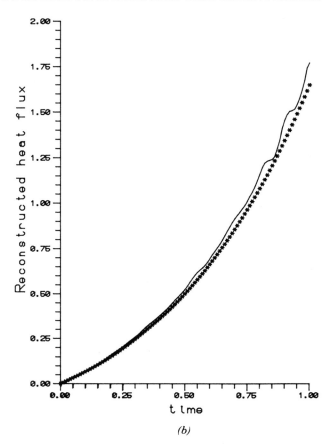

(b)

Fig. 4.2b   Reconstructed heat flux function—Problem 2: $x = y = 1/2$, $\varepsilon = 0.005$, $\delta = 0.04$, $h = 0.1$, $s = 0.015$, $k = 0.01$. Exact values ($* * *$); computed solution (—).

with corresponding approximate data function $F_m(y, t)$,

$$u_x(0, y, t) = Q(y, t), \qquad 0 < y < 1, t > 0, \tag{4.19c}$$

with corresponding approximate data function $Q_m(y, t)$,

$$u(x, y, 0) = 0, \qquad 0 \le x \le 1/2, 0 < y < 1, \tag{4.19d}$$

$$u(1/2, y, t) = f(y, t), \qquad 0 < y < 1, t > 0, \tag{4.19e}$$

the desired but unknown temperature function,

$$u_x(1/2, y, t) = q(y, t), \qquad 0 < y < 1, t > 0, \tag{4.19f}$$

the desired but unknown heat flux function,

$$u(x, 0, t) = h_0(x, t), \qquad 0 \le x \le 1/2, t > 0, \qquad (4.19\text{g})$$

unknown boundary temperature function,

$$u(x, 1, t) = h_1(x, t), \qquad 0 \le x \le 1/2, t > 0, \qquad (4.19\text{h})$$

unknown boundary temperature function.

A few words about the uniqueness of the solution of system (4.19) are appropriate.

To that end, consider the direct problem for the two-dimensional heat conduction system given previously, with boundary data temperature $F(y, t) = u(0, y, t)$, $\bar{h}_0(x, t) = u(x, 0, t)$, $\bar{h}_1(x, t) = u(x, 1, t)$, and $\bar{f}(y, t) = u(1/2, y, t)$. This problem has a unique temperature solution and, in particular, an induced heat flux solution at $x = 0$ that we denote $\bar{Q}(y, t) = u_x(0, y, t)$. On the other hand, the same direct problem with data temperature $F(y, t) = v(0, y, t)$, $h_0(x, t) = v(x, 0, t)$, $h_1(x, t) = v(x, 1, t)$, and $f(y, t) = v(1/2, y, t)$ also has a unique solution and the induced heat flux at $x = 0$ will be denoted $Q(y, t) = v_x(0, y, t)$.

Is it possible to have $Q = \bar{Q}$ if $f \neq \bar{f}$ and/or $h_0 \neq \bar{h}_0$ and/or $h_1 \neq \bar{h}_1$? By superposition, the solution of the direct problem with boundary data temperature $\Delta\bar{F} = F - \bar{F} = 0$, $\Delta h_0 = h_0 - \bar{h}_0$, $\Delta h_1 = h_1 - \bar{h}_1$, and $\Delta f = f - \bar{f}$ is such that the heat flux at $x = 1/2$, $\Delta\tilde{Q} = Q - \bar{Q}$, is identically 0. We can now imbed the original rectangular domain symmetrically into the square $[1/2, 1/2] \times [0, 1]$ in the (x, y) plane—for every time t—and seek some relationship (if any) among the boundary data functions Δf, Δh_0, Δh_1, ΔF, ΔH_0, and ΔH_1 such that $\Delta\bar{F} = \Delta\tilde{Q} = 0$. Pictorially, with $T = u - v$, the situation looks like this:

$$
\begin{array}{ccc}
 & \Delta H_1 & \qquad\qquad \Delta h_1 \\
\end{array}
$$

$$
\Delta F \quad
\begin{array}{|c|c|}
\hline
\begin{array}{l} \Delta\tilde{F} = \\ T_t = T_{xx} + T_{yy} \\ \Delta\tilde{Q} = \end{array} & \begin{array}{l} 0 \\ T_t = T_{xx} + T_{yy} \\ 0 \end{array} \\
\hline
\end{array}
\quad \Delta f
$$

$$
\begin{array}{ccc}
 & \Delta H_0 & \qquad\qquad \Delta h_0 \\
\end{array}
$$

Given that we do not consider interior sources in our heat transfer model, the condition $\Delta\bar{F} = 0$ can be achieved if and only if "the effect of each of the two symmetric boundary data temperatures annihilates each other at every point (x, y, t)," that is, if and only if

$$\Delta f = -\Delta F, \qquad \Delta h_1 = -\Delta H_1, \qquad \text{and} \qquad \Delta h_0 = -\Delta H_0. \quad (4.20)$$

The same argument implies that for the heat flux condition $\Delta \tilde{Q} = 0$ to occur, we must have

$$\Delta f = \Delta F, \qquad \Delta h_1 = \Delta H_1, \qquad \text{and} \qquad \Delta h_0 = \Delta H_0. \qquad (4.21)$$

From equalities (4.20) and (4.21), it follows that

$$\Delta f = \Delta h_1 = \Delta h_0 = 0,$$

and we conclude that $\Delta \tilde{F} = \Delta \tilde{Q} = 0$ if and only if the temperature data on the boundary is identically 0. ∎

This argument shows that the IHCP (4.19) has a unique solution.

We once again approximate the system (4.19) with the consistent finite-difference scheme

$$W_{i+1,j}^n = W_{i,j}^n + \frac{h}{2k}\left(V_{i,j}^{n+1} - V_{i,j}^{n-1}\right) - \frac{h}{s^2}\left(V_{i,j-1}^n - 2V_{i,j}^n + V_{i,j+1}^n\right),$$

$$V_{i+1,j}^n = V_{i,j}^n + hW_{i+1,j}^n,$$

$$i = 0, 1, \ldots, N-1, \, j = 1, 2, \ldots, M-1, \, n = 1, 2, \ldots, P-i-1,$$

$$V_{0,j}^n = \left(J_\delta F_m\right)_j^n, \qquad j = 0, 1, 2, \ldots, M, \, n = 1, 2, \ldots, P-1,$$

$$W_{0,j}^n = \left(J_\delta Q_m\right)_j^n, \qquad j = 0, 1, 2, \ldots, M, \, n = 1, 2, \ldots, P-1,$$

$$V_{i,j}^0 = 0, \qquad i = 1, 2, \ldots, N, \, j = 1, 2, \ldots, M.$$

As we march forward in the x direction in space, we still drop the estimation of the interior temperature from the highest previous point in time but the discrete heat fluxes $W_{i+1,0}^n$ and $W_{i+1,M}^n$ are computed by linear extrapolation from the already estimated values $W_{i+1,1}^n$, $W_{i+1,2}^n$ and $W_{i+1,M-2}^n$, $W_{i+1,M-1}^n$, introducing an extra local error of order $O(s)$:

$$W_{i+1,0}^n = 2W_{i+1,1}^n - W_{i+1,2}^n,$$

$$W_{i+1,M}^n = 2W_{i+1,M-1}^n - W_{i+1,M-2}^n.$$

The corresponding unknown boundary temperatures are then calculated directly using the usual formulas

$$V_{i+1,0}^n = V_{i,0}^n + hW_{i+1,0}^n$$

and

$$V_{i+1,M}^n = V_{i,M}^n + hW_{i+1,M}^n.$$

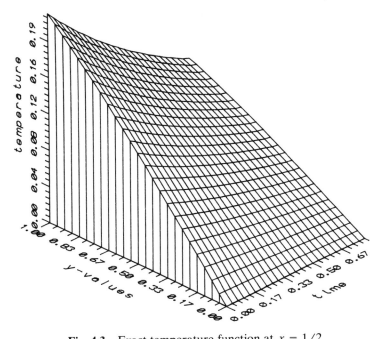

Fig. 4.3 Exact temperature function at $x = 1/2$.

This process is continued for $n = 1, 2, \ldots, P - i - 1$ until the approximate solution is computed at all the grid points of the domain in the plane x_{i+1}. Then the entire cycle is repeated for the discrete points of the plane x_{i+2}, and so on.

Figures 4.3 to 4.5 show the exact temperature solution restricted to the grid points, the discrete computed solution—obtained with the numerical scheme introduced previously with $h = 0.1$, $s = k = 0.01$, $\delta = 0.04$, $\varepsilon = 0.005$—and the corresponding discrete error surface at $x = 1/2$ and $0 \le t \le 0.8$, respectively, for the following IHCP:

$$
\begin{aligned}
u_t &= u_{xx} + u_{yy}, & & 0 < x < 1/2, 0 < y < 1, t > 0, \\
u(0, y, t) &= 0, & & 0 < y < 1, t > 0, \\
u_x(0, y, t) &= \frac{1}{\sqrt{2}} e^{-t} \sin\frac{y}{\sqrt{2}}, & & 0 < y < 1, t > 0, \\
u(x, y, 0) &= \sin\frac{y}{\sqrt{2}} \sin\frac{x}{\sqrt{2}}, & & 0 \le x \le 1/2, 0 < y < 1, \quad (4.22) \\
u(1/2, y, t) &= f(y, t), \text{ unknown}, & & 0 < y < 1, t > 0, \\
u_x(1/2, y, t) &= q(y, t), \text{ unknown}, & & 0 < y < 1, t > 0, \\
u(x, 0, t) &= h_0(x, t), \text{ unknown}, & & 0 \le x \le 1/2, t > 0, \\
u(x, 1, t) &= h_1(x, t), \text{ unknown}, & & 0 \le x \le 1/2, t > 0.
\end{aligned}
$$

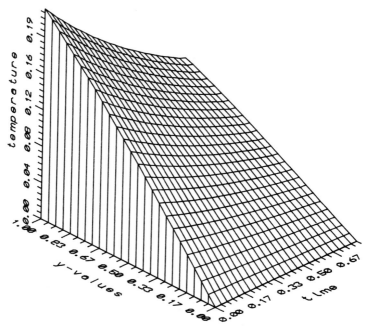

Fig. 4.4 Reconstructed temperature function at $x = 1/2$: $h = 0.1$, $s = k = 0.01$, $\delta = 0.04$, $\varepsilon = 0.005$.

The unique exact temperature solution for problem (4.22) is

$$u(x, y, t) = e^{-t} \sin \frac{y}{\sqrt{2}} \sin \frac{x}{\sqrt{2}}, \qquad 0 \le x \le 1/2, 0 \le y \le 1, t > 0,$$

and, consequently, the unknown functions for the IHCP are given by

$$f(y, t) = e^{-t} \sin \frac{y}{\sqrt{2}} \sin \frac{1}{2\sqrt{2}}, \qquad 0 \le y \le 1, t > 0,$$

$$q(y, t) = \frac{1}{\sqrt{2}} e^{-t} \sin \frac{y}{\sqrt{2}} \cos \frac{1}{2\sqrt{2}}, \qquad 0 \le y \le 1, t > 0,$$

$$h_0(x, t) = h_0(x, t) = 0, \qquad 0 \le x \le 1/2, t > 0,$$

$$h_1(x, t) = e^{-t} \sin \frac{x}{\sqrt{2}} \sin \frac{1}{\sqrt{2}}, \qquad 0 \le x \le 1/2, t > 0.$$

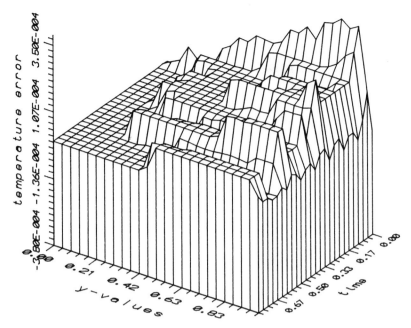

Fig. 4.5 Temperature error function at $x = 1/2$: $h = 0.1$, $s = k = 0.01$, $\delta = 0.04$, $\varepsilon = 0.005$.

4.5   EXERCISES

4.1. Show that given $\varepsilon > 0$, there exists a unique radius of mollification $\delta > 0$, such that

$$\|J_\delta F - F\|_D = \varepsilon, \tag{4.23}$$

where $\| \cdot \|_D$ indicates the L^2 norm in the unit square $D = [0, 1] \times [0, 1]$ in the (y, t) plane and $J_\delta F$ represents the two-dimensional convolution of the Gaussian kernel $\rho_\delta(y, t)$ with the locally integrable function $F(t, y)$. Here

$$\rho_\delta(t, y) = \frac{1}{\pi \delta^2} \exp\left[-\frac{y^2 + t^2}{\delta^2} \right].$$

4.2. Discuss at least two possible implementations for solving (4.23) in Exercise 4.1.

4.3. If the two-dimensional convolution in Exercise 4.1 is replaced by two successive one-dimensional convolutions in t and y, how can we accelerate the data filtering process? Should we modify the selection criterion of the radius of mollification in this case?

4.4. Show that the uniqueness result for the two-dimensional IHCP on a bounded rectangular domain—and all times t—(4.19) holds in the more general case of a bounded domain $\Omega \subset \mathbb{R}^2$ with smooth boundary $\partial\Omega$ if the temperature and the normal heat flux are specified in a portion of $\partial\Omega \times [0, t)$.

4.6 REFERENCES AND COMMENTS

The following references and comments serve to expand the basic material covered in the corresponding sections.

The first analytical solution—requiring exact data—to the inverse heat conduction problem that is applicable to two-dimensional conduction systems for geometries of arbitrary shape was presented in

M. Imber, Temperature extrapolation mechanism for two-dimensional heat flow, *AIAA J.* **12** (1974), 1089–1093.

4.1. Most of the literature related to the two-dimensional IHCP is based on different ways of combining finite-element realizations with the future temperatures method introduced in Chap. 3. For some of the early applications of these ideas, see

B. R. Bass and L. J. Ott, A finite element formulation of the two-dimensional inverse heat conduction problem, *Adv. in Comput. Tech.* **2** (1980), 238–248.

More numerical experimentation can be found in

T. Yoshimura and K. Ituka, Inverse heat conduction problem by finite element formulation, *Internat. J. Systems Sci.* **16** (1985) 1365–1376.

A comprehensive exposition was presented later in

J. Baumeister and H. J. Reinhart, On the approximate solution of a two-dimensional inverse heat conduction problem, *Inverse and Ill-Posed Problems*, H. G. Engl and C. W. Groetsch (eds.), Academic, Orlando, 1987, pp. 325–344.

N. Zabaras and J. C. Liu, An analysis of two-dimensional linear inverse heat transfer problems, using an integral method, *Numer. Heat Transfer* **13** (1988) 527–533.

4.2–4.4. Here we follow in part the work of

L. Guo and D. A. Murio, A mollified space-marching finite difference algorithm for the two-dimensional inverse heat conduction problem with slab symmetry, *Inverse Problems* **7** (1991) 247–259.

5

APPLICATIONS OF THE SPACE MARCHING SOLUTION OF THE IHCP

There are several different physical situations where the space marching approach for the solution of the IHCP allows for a more general and natural point of view of the original problems, partially eliminating the sharp dichotomy between the mathematical treatment of ill-posed and well-posed tasks.

For instance, in the identification of boundary source functions and nonlinear boundary radiation laws, the original well-posed problem is replaced by a related stabilized ill-posed one—a daring suggestion—gaining great advantages in terms of location and collection of the data necessary to solve the problem. This is also the case in other applications related to solidification phenomena in phase-change materials, when the evolution of the solid–liquid interface is sought (Stefan problem). The space marching scheme can also be quite naturally applied—with no significant changes—to the direct solution of the ill-posed inverse Stefan problem where the front evolution is approximately known and one wants to find the temperature and heat flux histories at a fixed liquid boundary location. Finally, in another quite different application, we suggest how to approximately recover the initial temperature distribution in the one-dimensional IHCP while marching in space, illustrating an interesting relationship between the backward-in-time heat conduction problem and the IHCP.

5.1 IDENTIFICATION OF BOUNDARY SOURCE FUNCTIONS

In this section we investigate the numerical identification of surface transient heat sources in one-dimensional semi-infinite and finite slab mediums when the active surface radiates energy according to a known nonlinear law. Alternatively, if the active surface is heated by a source at a rate proportional to a given function, the nonlinear radiating boundary condition is then numerically identified as a function of the interface temperature.

These two tasks can be viewed as suitable generalizations of the classical problem of attempting to determine the interface temperature between a gas and a solid with a nonlinear heat transfer law. Indeed, the nonlinear radiation law and the transient boundary source are supposed to be known in order to determine the interface temperature. Consequently, if the new task consists of the identification of the transient boundary source function, a different approach must be used.

We consider a one-dimensional IHCP in a semi-infinite or finite slab, in which the temperature and heat flux histories $f(t)$ and $q(t)$ on the left-hand surface $(x = 0)$ are desired and unknown, and the temperature and heat flux at some interior point $x = x_0$ or on the right-hand surface $(x = 1)$ are approximately measurable. Note that, equivalently, the data temperature histories might be measured at two interior points. For the semi-infinite medium, $0 < x_0$, and for the finite slab, $0 < x \leq 1$. We assume linear heat conduction with constant coefficients and normalize the problem by dimensionless quantities. Without loss of generality, we consider $x_0 = 1$ in all cases. The problem can be described mathematically as follows.

For the semi-infinite or finite slab, the unknown temperature $u(x, t)$ satisfies, respectively,

$$u_t(x,t) = u_{xx}(x,t), \qquad t > 0, 0 < x < \infty \text{ or } 0 < x < 1, \qquad (5.1a)$$

$$u(1,t) = F(t), \qquad t > 0, \qquad (5.1b)$$

with corresponding approximate data function $F_m(t)$,

$$-u_x(1,t) = Q(t), \qquad t > 0, \qquad (5.1c)$$

with corresponding approximate data function $Q_m(t)$,

$$u(x,0) = u_0(x), \qquad 0 < x < \infty \text{ or } 0 < x < 1, \qquad (5.1d)$$

$$u(0,t) = f(t), \qquad t > 0, \qquad (5.1e)$$

the desired but unknown temperature function,

$$-u_x(0,t) = q(t)$$
$$= E(u(0,t)) - G(t), \qquad t > 0, \qquad (5.1f)$$

the desired but unknown heat flux function.

The nonlinear boundary condition (5.1f) indicates that the active surface radiates energy at a rate proportional to E and is heated at a rate proportional to the function G. Our aim is to obtain more detailed information about the boundary condition at the interface $x = 0$. More precisely, we want to estimate the function E if G is known or, reciprocally, we want to identify the source function G if the radiation law E is given.

As usual, we assume that all the functions involved are L^2 functions in any time interval of interest and hypothesize that the exact data functions $F(t)$ and $Q(t)$ and the measured data functions $F_m(t)$ and $Q_m(t)$ satisfy the L^2 data error bounds

$$\|F - F_m\| \le \varepsilon \qquad \text{and} \qquad \|Q - Q_m\| \le \varepsilon.$$

We recall that under the conditions stated in Theorem 3.4, the temperature and heat flux boundary functions—corresponding to the mollified problem associated with (5.1)—satisfy on the bounded time interval $D = [0, T]$:

$$\|f - J_\delta f_m\|_D \le \tfrac{1}{2}M\delta + \varepsilon \exp\left[(2\delta)^{-2/3}\right] \tag{5.2}$$

and

$$\|q - J_\delta q_m\|_D \le \tfrac{1}{2}M\delta + 2\varepsilon \exp\left[(2\delta)^{-2/3}\right]. \tag{5.3}$$

Once the mollified temperature and mollified heat flux functions have been evaluated at the interface, it is feasible to attempt to identify the source function G or the radiation energy function E given in formula (5.1f).

Identification of the Source Function G
Assuming that the radiation law at the active surface is known, according to (5.1f), the exact source function is given by

$$G(t) = E(f(t)) - q(t). \tag{5.4}$$

The approximate source function, denoted $G_a(t)$, is defined by

$$G_a(t) = E(J_\delta f_m(t)) - J_\delta q_m(t), \tag{5.5}$$

and in order to estimate the error, we suppose that the surface radiates energy at a rate proportional to $[f(t)]^p$. Here p is a positive integer, the value $p = 1$ corresponding to Newton's law of cooling, and $p = 4$ to Stefan's radiation law.

The difference (5.4) − (5.5) gives

$$G(t) - G_a(t) = [f(t)]^p - [J_\delta f_m(t)]^p - q(t) + J_\delta q_m(t).$$

From the identity $x^n - b^n = (x - b)(x^{n-1} + x^{n-2}b + \cdots + xb^{n-2} + b^{n-1})$, taking norms and introducing $M_\delta = \max\{\|J_\delta f_m\|_{\infty, D}, \|f\|_{\infty, D}\}$, we get

$$\|G - G_a\|_D \le pM_\delta^{p-1}\|f - J_\delta f_m\|_D + \|q - J_\delta q_m\|_D.$$

Combining the last inequality with the upper bounds (5.2) and (5.3), we obtain the estimate

$$\|G - G_a\|_D \leq (pM_\delta^{p-1} - 1)\left\{\tfrac{1}{2}M\delta + 2\varepsilon \exp\left[(2\delta)^{-2/3}\right]\right\}. \tag{5.6}$$

This shows that the identification of the source function G is stable with respect to errors in the data functions F and Q, for fixed p and $\delta > 0$.

Remarks

1. Notice that the approximate source function G_a is actually a function of the radius of mollification δ, the amount of noise in the data ε, and the exponent p in the radiation model E.
2. From a more theoretical point of view, inequality (5.6) can be used to show the convergence of G_a to G in the L_2 norm for a suitable choice of the radius of mollification δ. In fact, setting $\delta = [\ln(1/\varepsilon^{1/2})]^{-3/2}$, after replacing this quantity in (5.6), we obtain

$$\|G - G_a\|_D \leq (pM_\delta^{p-1} + 1)\left\{\tfrac{1}{2}M\left[\ln(1/\varepsilon^{1/2})\right]^{-3/2} + 2\varepsilon^{3/4}\right\}.$$

This last inequality implies that, for the special selection of the radius of mollification indicated previously, $\|G - G_a\|_D \to 0$ as $\varepsilon \to 0$, for any value of p.

Identification of the Radiation Law Function E
From (5.1f) it follows that the exact function E, assuming that the source function G is given, satisfies

$$E(u(0, t)) = E(f(t)) = G(t) + q(t). \tag{5.7}$$

The approximate law, denoted E_a, is defined by

$$E_a(J_\delta f_m(t)) = G(t) + J_\delta q_m(t). \tag{5.8}$$

Subtracting (5.7) from (5.8), taking norms, and using inequality (5.3), we immediately have

$$\|E - E_a\|_D \leq \tfrac{1}{2}M\delta + 2\varepsilon \exp\left[(2\delta)^{-2/3}\right]. \tag{5.9}$$

This estimate also shows that the identification of the radiation law—*as a function of time*—is stable with respect to perturbations in the data functions F and Q, for a fixed $\delta > 0$, provided that the source function is known. However, this information is clearly not sufficient to identify the physical process at the interface. Nevertheless, since at each time t_i we know the ordered pairs $(t_i, J_\delta f_m(t_i))$ and $(t_i, E_a(t_i))$, it is possible to collect the coordinates $(J_\delta f_m(t_i), E_a(t_i))$ for t in a discrete subset of D and obtain a graph of the approximate functional relationship between the radiation law and the temperature at the interface. This is certainly always the case if the cardinality of the range of temperatures $\{J_\delta f_m t(i)\}$ is sufficiently large. Similar remarks to the ones in the previous paragraph, about the parameter dependency of E_a and convergence in the L_2 norm of E_a to E as the quality of the data functions improve, $\varepsilon \to 0$, also apply here.

The computational details are presented in the next section.

5.2   NUMERICAL PROCEDURE

We approximate the mollified partial differential equation associated with system (5.1) with the consistent finite-difference scheme (3.47) introduced in Chap. 3:

$$W_{i-1}^n = W_i^n - \frac{h}{2k}(V_i^{n+1} - V_i^{n-1}),$$

$$V_{i-1}^n = V_i^n - hW_{i-1}^n, \tag{5.10}$$

$$i = N, N-1, \ldots, 1, n = 1, 2, \ldots, M-1.$$

Notice that $V_i(x_i, t_n) = J_\delta u(x_i, t_n)$, as before, but

$$W_i(x_i, t_n) = -\partial/\partial x J_\delta u(x_i, t_n)$$

instead of $\partial/\partial x J_\delta u(x_i, t_n)$. Also,

$$V_N^n = J_\delta F_m(t_n), \qquad W_N^n = J_\delta Q_m(t_n), \qquad 0 \le n \le M,$$

and

$$V_i^0 = u_0(x_i), \qquad 0 \le i \le N.$$

Once the temperature $J_\delta f_m$ and the heat flux $J_\delta q_m$ have been reconstructed, we proceed with the approximate identification of the source function G_a or the radiation law function E_a as explained in Sec. 5.1.

Numerical Results

In Problem 1 the approximate reconstruction of a source function $G(t)$ and a nonlinear radiation law $E(u(0, t))$ are investigated for a one-dimensional finite slab exposed to a heat flux data function at the free surface $x = 1$ given by

$$-u_x(1, t) = Q(t) = 0, \qquad t > 0,$$

and a temperature data function

$$u(1, t) = F(T)$$

$$= \begin{cases} (t - 0.2) - \dfrac{1}{6} - \dfrac{2}{\pi^2} \displaystyle\sum_{n=1}^{\infty} \dfrac{(-1)^n}{n^2} \exp[-n^2\pi^2(t - 0.2)], & t > 0.2, \\ 0, & 0 < t \leq 0.2. \end{cases}$$

The exact source solution to be approximately reconstructed at the interface $x = 0$ has equation $G(t) = E(u(0, t)) - q(t)$, where $E(u(0, t)) = [u(0, t)]^p$ and $q(t) = -u_x(0, t)$. We consider the values $p = 1$ and $p = 4$ corresponding to Newton's law of cooling and Stefan's radiation law, respectively. The exact radiation law at the interface is given by $E(u(0, t)) = G(t) + q(t)$ and we only consider the nonlinear case $p = 4$. If the initial temperature distribution $u(x, 0)$ is 0, the exact temperature and heat flux functions at the interface are given respectively by

$$u(0, t) = f(t)$$

$$= \begin{cases} (t - 0.2) + \dfrac{1}{3} - \dfrac{2}{\pi^2} \displaystyle\sum_{n=1}^{\infty} \dfrac{1}{n^2} \exp[-n^2\pi^2(t - 0.2)], & t > 0.2, \\ 0, & 0 < t \leq 0.2, \end{cases}$$

and

$$-u_x(0, t) = q(t) = \begin{cases} 1, & t > 0.2, \\ 0, & 0 < t \leq 0.2, \end{cases}$$

With this information we generate the exact functions $E(u(0, t))$ and $G(t)$ for our model problem.

In Problem 2 we attempt to approximately reconstruct the transient source function $G(t)$ for a semi-infinite body initially at zero temperature with data functions

$$u(1, t) = F(t) = \begin{cases} \text{erfc}\left[\dfrac{1}{2(t - 0.2)^{1/2}} \right], & t > 0.2, \\ 0, & 0 < t \leq 0.2, \end{cases}$$

and

$$-u_x(1, t) = Q(t)$$

$$= \begin{cases} [\pi(t - 0.2)]^{-1/2} \exp\{-[4(t - 0.2)]^{-1}\}, & t < 0.2, \\ 0, & 0 < t \leq 0.2. \end{cases}$$

The unique temperature solution at the interface is

$$u(0, t) = f(t) = \begin{cases} 1, & t > 0.2, \\ 0, & 0 < t \le 0.2, \end{cases}$$

and the corresponding heat flux at the interface is

$$-u_x(0, t) = q(t) = \begin{cases} [\pi(t - 0.2)]^{-1/2}, & t > 0.2, \\ 0, & 0 < t \le 0.2. \end{cases}$$

In this case, we do not attempt the identification of the radiation law at the active boundary. The energy as a function of the interface temperature is either 0 or 1 for any value of p making its identification impossible. There is not enough information in the range of boundary temperatures which in this example is reduced to just two temperature values.

Once the radii of mollification δ_F and δ_Q associated with the data functions F_m and Q_m, respectively, and the discrete filtered data functions $J_\delta F_m(t_n) = V_N^n$ and $J_\delta Q_m(t_n) = W_N^n$, $0 \le n \le M$, are determined with $\delta = \max(\delta_F, \delta_Q)$, we apply the finite-difference algorithm described previously in this section, marching backward in the x direction. The values V_0^n and W_0^n, $0 \le n \le M - N$, so obtained, are then taken as the accepted approximations for the interface temperature and heat flux histories, respectively, at the different time locations at $x = 0$. Finally, we identify the approximate transient source function G_a or the approximate radiation law function E_a at the grid points of the time interval $I = [0, 1]$ using (5.5) and (5.8).

In all cases, we use $h = \Delta x = 0.01$ and $k = \Delta t = 0.01$. The noisy data are obtained, as in the previous models, by adding a random error to the exact discrete data function at every grid point of the extended data interval.

If the discretized computed transient source function component is denoted by G_a^n and the true component is $G^n = G(t_n)$, the weighted l^2 norm of the solution error is then given, in the time interval $I = [0, 1]$, by

$$\|G_a - G\|_I = \left[\frac{1}{M - N} \sum_{n=1}^{M-N} (G_a^n - G^n)^2 \right]^{1/2}.$$

TABLE 5.1. Error Norm as a Function of the Level of Noise in Problem 1

	$p = 1$ (Newton)			$p = 4$ (Stefan)	
ε	δ	Error Norm	ε	δ	Error Norm
0.000	0.04	0.0866	0.000	0.04	0.0867
0.002	0.06	0.0921	0.002	0.06	0.0929
0.005	0.06	0.1014	0.005	0.06	0.1038

TABLE 5.2. Error Norm as a Function of the Level of Noise in Problem 2

	$p = 1$ (Newton)			$p = 4$ (Stefan)	
		Error Norm			Error Norm
ε	δ	$[0, 1]/[0.3, 1]$	ε	δ	$[0, 1]/[0.3, 1]$
0.000	0.04	0.5208/0.0183	0.000	0.04	0.5135/0.0373
0.002	0.06	0.5560/0.0631	0.002	0.06	0.5673/0.0675
0.005	0.06	0.5879/0.1108	0.005	0.06	0.5935/0.1375

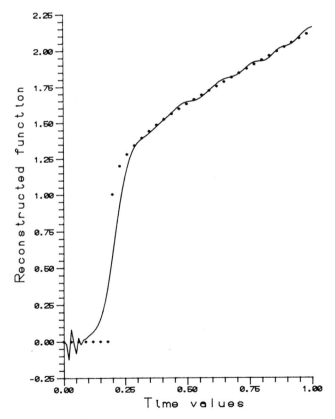

Fig. 5.1 Reconstructed boundary source function. Newton's law—Problem 1: $\varepsilon = 0.005$, $\delta = 0.06$, $h = k = 0.01$. Exact ($\cdots$); computed (—).

If the discretized computed radiation law function component is denoted by $E_a^n = E_a(t_n)$ and the true component is $E^n = E(t_n)$, after evaluating the ordered pairs (V_0^n, E_a^n), $0 \leq n \leq M - N$, we obtain a graph of the approximate functional relationship between the radiation law and the temperature at the interface. This plot is then compared with the exact graph corresponding to the values $(f(t_n), E(t_n))$ of the model problem.

Tables 5.1 and 5.2 show the results of our numerical experiments associated with Problems 1 and 2, respectively, when attempting to identify the transient source function at the interface. The uniformly smaller error norms in Problem 1 are expected since at time $t = 0.2$ the exact source solution has a finite-jump discontinuity whereas in Problem 2 the exact source solution has an infinite jump at time $t = 0.2$. For this reason, we have added an extra column in Table 5.2 indicating the error norms in the time interval $[0.3, 1]$, after the discontinuity. It is clear that the method rapidly dissipates the effect of the singularity, a very desirable feature.

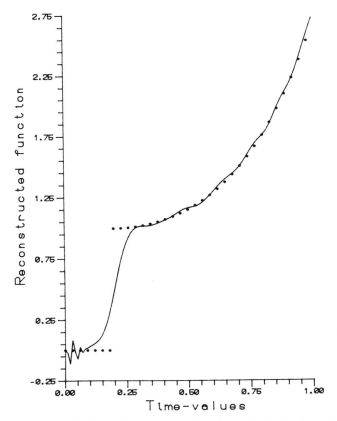

Fig. 5.2 Reconstructed boundary source function. Stefan's law—Problem 1: $\varepsilon = 0.005$, $\delta = 0.06$, $h = k = 0.01$. Exact ($\cdots$); computed (—).

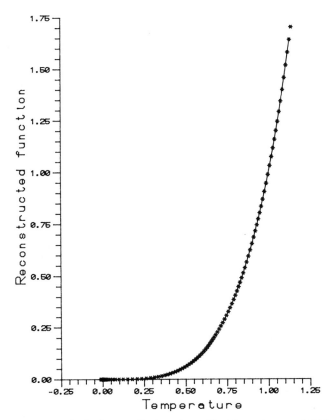

Fig. 5.3 Reconstructed Stefan radiation law. Problem 1: $\varepsilon = 0.005$, $\delta = 0.06$, $h = k = 0.01$. Exact (——); computed ($* * *$).

The qualitative behavior of the reconstructed transient source function for Problem 1 is illustrated in Figs. 5.1 and 5.2 where the numerical solution for an average perturbation $\varepsilon = 0.005$ (full line) is plotted for $p = 1$ (Newton's cooling law) and $p = 4$ (Stefan's radiation law), respectively. In Fig. 5.3 we show the graph associated with the reconstructed nonlinear radiation law as a function of the approximate temperature at the interface for $p = 4$ (starred symbols) and the exact boundary radiation law (full line). Figures 5.4 and 5.5 show the computed source functions (full lines) for $p = 1$ and $p = 4$, respectively, for Problem 2 and for the noise level $\varepsilon = 0.005$.

In both problems the source functions to be identified have discontinuous histories and in one case an infinite jump. The algorithm restores stability with respect to the data, which is essential for the introduction of the inverse problem approach, and good accuracy is obtained, even for small time sample intervals and relative high noise levels in the data.

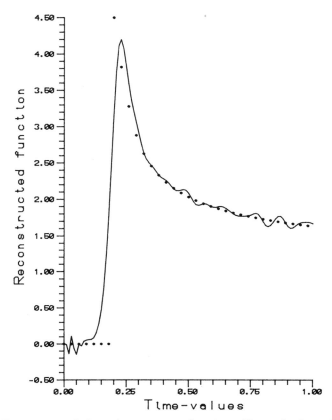

Fig. 5.4 Reconstructed boundary source function. Newton's law—Problem 2: $\varepsilon = 0.005$, $\delta = 0.06$, $h = k = 0.01$. Exact ($\cdots$); computed (—).

5.3 IHCP WITH PHASE CHANGES

In many applications related to solidification phenomena in phase-change materials, it is important to know with good accuracy the evolution of the solid–liquid interface.

It is possible to consider suitable inverse heat conduction problems defined only on the liquid domain or on the solid domain to approximately determine the transient location of the interface. In this setting, the one-dimensional Stefan problem is simply considered as a one-dimensional IHCP with an unknown moving boundary location which is assumed to be an isotherm. To solve the associated IHCP, the temperature and heat flux history functions are needed at one location of the liquid (solid) domain or, alternatively, two temperature histories at two different locations of the liquid (solid) domain must be measured. This point of view amounts to

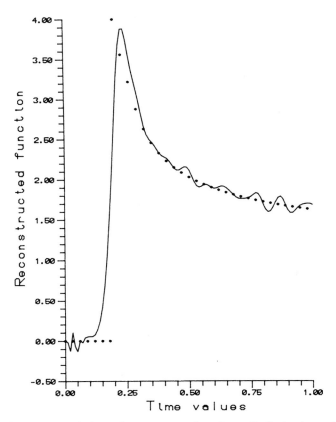

Fig. 5.5 Reconstructed boundary source function. Stefan's law—Problem 2: $\varepsilon = 0.005$, $\delta = 0.06$, $h = k = 0.01$. Exact ($\cdots$); computed (—).

replacing Stefan's well-posed problem by the IHCP, an inherently ill-posed problem: small errors in the input data might induce large errors in the computed interface temperature which is needed to locate the front and, consequently, special methods are necessary in order to restore continuity with respect to the data.

We consider the implementation of the mollification method, originally developed for the solution of the IHCP as a front-tracking space marching procedure to determine the location of the solid–liquid interface in one-dimensional melting problems.

The space marching method is also applied to the solution of the inverse Stefan problem: if the front evolution and the heat flux at the melting interface are approximately known, the task consists now of finding the temperature and heat flux histories at the fixed liquid (solid) boundary location. It is quite natural to attempt to solve this problem with some variant of the IHCP since it is well known that the inverse Stefan problem is ill posed with respect to perturbations in the data.

5.4   DESCRIPTION OF THE PROBLEMS

Stefan Problem

Mathematically, the one-dimensional one-phase Stefan problem can be defined as follows. Assuming linear heat conduction with constant coefficients and dimensionless quantities, the unknown temperature in the liquid phase $u(x, t)$ satisfies

$$
\begin{aligned}
u_t(x, t) &= u_{xx}(x, t), & 0 &< x < s(t), 0 < t \le T, \\
u(x, 0) &= u_0(x), & 0 &\le x \le s(0), 0 < s(0), \\
u(0, t) &= F(t), & 0 &< t \le T, \\
u(s(t), t) &= 0, & 0 &< t \le T, \\
-u_x(s(t), t) &= \frac{ds}{dt}(t) = q(t), & 0 &< t \le T,
\end{aligned}
\tag{5.11}
$$

where the initial temperature distribution $u_0(x)$ and the fixed boundary temperature history $F(t)$ are known and one has to determine the transient interface location $s(t)$ and the heat flux at the front $q(t)$.

For the application of the IHCP techniques to the approximate solution of this problem, it is necessary to also measure the heat flux history at the fixed boundary $x = 0$ or, alternatively, to measure the temperature history at another location x_1, $0 < x_1 < s(0)$. Thus, we consider the problem

$$
u_t(x, t) = u_{xx}(x, t), \qquad 0 < x < s(t), 0 < t \le T, \tag{5.12a}
$$

$$
u(x, 0) = u_0(x), \qquad 0 \le x \le s(0) > 0, \tag{5.12b}
$$

$$
u(0, t) = F(t), \qquad 0 < t \le T, \tag{5.12c}
$$

with corresponding approximate data function $F_m(t)$,

$$
-u_x(0, t) = Q(t), \qquad 0 < t \le T, \tag{5.12d}
$$

with corresponding approximate data function $Q_m(t)$,

$$
u(s(t), t) = 0 = f(t), \qquad 0 < t \le T, \tag{5.12e}
$$

$$
-u_x(s(t), t) = \frac{ds}{dt}(t) = q(t), \qquad 0 < t \le T. \tag{5.12f}
$$

We also assume that all the functions involved are L_2 functions in any time interval D of interest and that the exact data functions $F(t)$ and $Q(t)$ and the measured data functions $F_m(t)$ and $Q_m(t)$ satisfy the L_2 data error bounds

$$\|F - F_m\|_D \le \varepsilon \qquad \text{and} \qquad \|Q - Q_m\|_D \le \varepsilon.$$

Stabilized Problem

Liquid Phase
Mollifying system (5.12), we obtain the following associated problem: attempt to find $J_\delta f_m(t) = J_\delta u(s_\delta(t), t)$ and $J_\delta q_m(t) = -J_\delta u_x(s_\delta(t), t)$ at some point t of interest and for some radius $\delta > 0$, given that $J_\delta u(x, t)$ satisfies in the liquid phase

$$
\begin{aligned}
(J_\delta u)_t &= (J_\delta u)_{xx}, & 0 < x < s_\delta(t), 0 < t \le T, \\
J_\delta u(x, 0) &= J_\delta u_0(x, 0), & 0 \le x \le s_\delta(0) > 0, \\
J_\delta u(0, t) &= J_\delta F_m(t), & 0 < t \le T, \\
-J_\delta u_x(0, t) &= J_\delta Q_m(t), & 0 < t \le T, \\
J_\delta u(s_\delta(t), t) &= J_\delta f_m(t) = 0, & 0 < t \le T, \\
-J_\delta u_x(s_\delta(t), t) &= J_\delta q_m(t), & 0 < t \le T.
\end{aligned}
\tag{5.13}
$$

The normalized version of this problem and its solutions satisfy the following special version of Theorem 3.4.

Theorem 5.1 Suppose that $0 < s_\delta(0) < s_\delta(t) \le 1, 0 < t \le T, \|F - F_m\|_D \le \varepsilon$, and $\|Q - Q_m\|_D \le \varepsilon$. Then

(i) Problem (5.13) is a formally stable problem with respect to perturbations in the data.

(ii) If the exact boundary temperature function $f(t)$ and the exact heat flux function $q(t)$ have uniformly bounded first-order derivatives on the bounded domain $D = [0, T]$, then $J_\delta f_m$ and $J_\delta q_m$ verify

$$\|f - J_\delta f_m\|_D \le \tfrac{1}{2}M\delta + \varepsilon \exp\left[(2\delta)^{-2/3}\right] \tag{5.14}$$

and

$$\|q - J_\delta q_m\|_D \le \tfrac{1}{2}M\delta + 2\varepsilon \exp\left[(2\delta)^{-2/3}\right]. \tag{5.15}$$

Once the mollified temperature function has been evaluated, it is feasible to attempt to identify the interface function $s_\delta(t)$.

Identification of the Interface Function s(t)

In general, the recovered mollified temperature $J_\delta f_m$ will be 0 at the approximate isotherm $s_\delta(t)$ which is different from the exact isotherm $s(t)$. However, since the function $J_\delta u(s, t)$ is strictly increasing with respect to the first variable s for $0 < s(0) < s(T) \le 1$, under the condition $\|\partial J_\delta u / \partial s\|_D \ge k_\delta > 0$, it is possible to estimate the error $\|s_\delta - s\|_D$.

From

$$\|J_\delta u(s_\delta, \cdot) - J_\delta u(s, \cdot)\|_D \ge k_\delta \|s - s_\delta\|_D,$$

$$\|s_\delta - s\|_D \le (k_\delta)^{-1} \|J_\delta u(s_\delta, \cdot) - u(s, \cdot) + u(s, \cdot) - J_\delta u(s, \cdot)\|_D$$

$$= (k_\delta)^{-1} \|u(s, \cdot) - J_\delta u(s, \cdot)\|_D,$$

taking into account that $u(s, t) = 0$ and $J_\delta u(s_\delta, t) = 0$.

Hence, using the estimate (5.14) with

$$J_\delta u(s_\delta(t), t) = J_\delta f_m(t) \quad \text{and} \quad u(s(t), t) = f(t),$$

we get

$$\|s_\delta - s\|_D \le (k_\delta)^{-1} \left[\tfrac{1}{2} M\delta + \varepsilon \exp\{(2\delta)^{-2/3}\} \right]. \tag{5.16}$$

Remarks

1. Notice that the approximate interface function s_δ is actually a function of the radius of mollification δ and the amount of noise in the data ε.
2. Inequality (5.16) can be used to show the convergence of s_δ to s in the L_2 norm as in Remark 2 in Sec. 5.1.

Solid Phase

Clearly, it is also possible to state a similar problem for the solid phase: attempt to find $J_\delta f_m(t) = J_\delta u(s_\delta(t), t) = 0$ and $J_\delta q_m(t) = -J_\delta u_x(s_\delta(t), t)$ at some point of interest t and for some radius $\delta > 0$, given that $J_\delta u(x, t)$ satisfies in the solid phase

$$
\begin{aligned}
(J_\delta u)_t &= (J_\delta u)_{xx}, & s_\delta(t) &< x < 1, 0 < t \le T, \\
J_\delta u(x, 0) &= J_\delta u_0(x, 0), & s_\delta(0) &\le x \le 1, 0 \le s(0) < 1, \\
J_\delta u(1, t) &= J_\delta F_m(t), & 0 &< t \le T, \\
-J_\delta u_x(1, t) &= J_\delta Q_m(t), & 0 &< t \le T, \\
J_\delta u(s_\delta(t), t) &= J_\delta f_m(t) = 0, & 0 &< t \ge T, \\
-J_\delta u_x(s_\delta(t), t) &= J_\delta q_m(t), & 0 &< t \le T.
\end{aligned}
\tag{5.17}
$$

Remark. We notice that the same stability bounds developed for the "forward in space" problem in the liquid phase apply also to the solutions of the Stefan problem in the solid phase.

Inverse Stefan Problem

In the formulation of problem (5.11), assume that a given initial temperature distribution $u_0(x)$, a prescribed interface of zero temperature $s(t)$, and the transient heat flux at the front $-u_x(s(t), t) = ds/dt(t) = q(t)$ are known. The inverse Stefan problem consists on finding the temperature and heat flux history functions at the fixed boundary $x = 0$. This problem can be immediately reinterpreted in terms of the "backward in space" IHCP in the liquid phase, with an "initial" transient boundary.

Stabilized Problem

Proceeding as in the previous sections, we obtain the associated well-posed normalized problem: attempt to find $J_\delta F_m(t) = J_\delta u(0, t)$ and $J_\delta Q_m(t) = -J_\delta u_x(0, t)$ at some time t of interest and for some radius of mollification $\delta > 0$, given that $J_\delta u(x, t)$ satisfies in the liquid phase

$$
\begin{aligned}
(J_\delta u)_t &= (J_\delta u)_{xx}, & 0 < x < s(t) \leq 1, 0 < t \leq T, \\
J_\delta u(x, 0) &= J_\delta u_0(x, 0), & 0 \leq x < s(0), 0 < s(0) \leq 1, \\
J_\delta u(0, t) &= J_\delta F_m(t), & 0 < t \leq T, \text{ unknown,} \\
-J_\delta u_x(0, t) &= J_\delta Q_m(t), & 0 < t \leq T, \text{ unknown,} \\
J_\delta u(s(t), t) &= J_\delta f_m(t) = 0, & 0 < t \leq T, \\
-J_\delta u_x(s(t), t) &= J_\delta q_m(t), & 0 < t \leq T.
\end{aligned}
\tag{5.18}
$$

Remarks

1. The data function $q_m(t)$ represents the measured heat flux function at the interface and we assume $\|q_m - q\|_D \leq \varepsilon$.
2. The conclusions of Theorem 5.1 also apply here with the obvious changes in notation, that is, the corresponding estimates now read

$$
\|F - J_\delta F_m\|_D \leq \tfrac{1}{2} M\delta + \varepsilon \exp\left[(2\delta)^{-2/3}\right]
\tag{5.19}
$$

and

$$
\|Q - J_\delta Q_m\|_D \leq \tfrac{1}{2} M\delta + 2\varepsilon \exp\left[(2\delta)^{-2/3}\right].
\tag{5.20}
$$

The computational details are presented in the next section.

5.5 NUMERICAL PROCEDURE

We approximate the mollified partial differential equation associated with system (5.11) with the consistent finite-difference scheme (3.47), introduced in Chapter 3, with $V_i(x_i, t_n) = J_\delta u(x_i, t_n)$ and $W_i(x_i, t_n) = -\partial/\partial x J_\delta u(x_i, t_n)$, marching forward in the x direction.

At each space step, once the temperature V_i^n and the heat flux W_i^n have been reconstructed for $0 \le n \le n_{max}$, we proceed with the approximate identification of the front s_δ by first collecting the pair of indices $(i, n - 1)$, (i, n) for which $V_i^{n-1} \cdot V_i^n \le 0$, $1 \le n \le n_{max}$ (vertical search). Thus, for the space location ih, the time location of s_δ is obtained by linear interpolation. The horizontal search proceeds similarly: after collecting the pair of indices $(i - 1, n)$, (i, n) for which $V_{i-1}^n \cdot V_i^n \le 0$, $i \le n \le n_{max}$, for the time location ik, the space location of s_δ is then obtained by linear interpolation.

Remarks

1. While marching forward in space, the horizontal search for the location of the space coordinate of the interface S_δ at the time grid points requires the storage of the computed time vector temperature solution at the previous space location.

2. The implementation of the numerical procedure for the solution of the normalized Stefan problem in the solid domain proceeds as explained previously except that the data functions are given at the fixed location $x = 1$ and we march backward in space.

Numerical Results

Stefan Problem: Liquid Phase
In order to test the accuracy and the stability properties of our method, in Problem 3 the approximate reconstruction of an interface function $s_\delta(t)$ is investigated for a one-dimensional one-phase Stefan problem exposed to a heat flux data function at the liquid (fixed) surface $x = 0$ given by

$$-u(0, t) = Q(t) = 2^{-1/2} \exp(1 - 2^{-1/2} + t/2), \qquad t > 0,$$

and a temperature data function

$$u(0, t) = F(t) = \exp(1 - 2^{-1/2} + t/2) - 1, \qquad t \ge 0.$$

If the initial temperature distribution is

$$u(x, 0) = \exp(1 - 2^{-1/2} - 2^{-1/2}x) - 1, \qquad 0 \le x \le s(0),$$

then the exact temperature and heat flux functions in the liquid phase are given respectively by

$$u(x, t) = \exp(1 - 2^{-1/2} + t/2 - 2^{-1/2}x) - 1,$$
$$0 < x < s(t) \le 1, 0 < t \le T,$$

and

$$-u_x(x,t) = (u(x,t) + 1)2^{-1/2}, \qquad 0 < x < s(t) \le 1, 0 < t \le T.$$

The exact interface solution function to be approximately reconstructed has equation $s(t) = 2^{-1/2}t + 2^{1/2} - 1$ and the exact temperature and heat flux functions at the melting surface are, respectively, $u(s(t), t) = f(t) = 0$, $0 < t \le T$, and $-u_x(s(t), t) = q(t) = 2^{-1/2}$, $0 < t \le T$.

In Problem 4 we attempt to approximately reconstruct the transient interface function $s(t)$ and the heat flux $-u_x(s(t), t) = q(t)$ at the melting front for a normalized one-dimensional Stefan problem with exact data functions

$$u(0, t) = F(t) = -t - 0.5, \qquad 0 < t \le T,$$
$$-u_x(0, t) = Q(t) = -2, \qquad 0 < t \le T,$$

and initial temperature distribution, in the liquid domain, given by

$$u(x, 0) = -0.5x^2 + 2x - 0.5, \qquad 0 < x \le s(0).$$

The exact temperature solution in the liquid phase is

$$u(x, t) = -0.5x^2 + 2x - t - 0.5, \qquad 0 < x < s(t) \le 1, 0 < t \le T,$$

and the exact heat flux has equation

$$-u_x(x, t) = x - 2, \qquad 0 < x < s(t) \le 1, 0 < t \le T.$$

It follows that the exact transient interface location is given by $s(t) = 2 - (3 - 2t)^{1/2}$, $0 \le t \le T$, and the temperature and heat flux solutions at the liquid–solid interface are, respectively, $u(s(t), t) = f(t) = 0$, $0 < t \le T$, and $-u_x(s(t), t) = q(t) = -(3 - 2t)^{1/2}$, $0 < t \le T$.

Once the radii of mollification δ_F and δ_Q associated with the data functions F_m and Q_m, respectively, and the discrete filtered data functions $J_\delta F_m(t_n) = V_0^n$ and $J_\delta Q_m(t_n) = W_N^n$, $0 \le n \le M$, are determined with $\delta = \max(\delta_F, \delta_Q)$, we apply the finite-difference algorithm described previously in this section, marching forward in the x direction and activating the horizontal and vertical search techniques, as explained before, to approximately determine the location of the melting front. The coordinates (x_δ, t_δ) so obtained are then taken as the accepted approximations for the interface location. Finally, the discrete heat flux history $q(x_\delta, t_\delta)$ is evaluated at those coordinates by linear interpolation.

In all cases, we use $h = \Delta x = 0.01$ and $k = \Delta t = 0.01$. The noisy data are obtained by adding a random error to the exact discrete data function at every grid point of the extended data interval.

If the discretized computed transient front function component is denoted by s_δ^n and the true component is $s^n = s(t_n)$, the weighted l^2 norm of the solution error is then given by

$$\|s_\delta - s\|_D = \left[\frac{1}{L} \sum_{n=1}^{L} (s_\delta^n - s^n)^2 \right]^{1/2}, \tag{5.21}$$

where L indicates the number of points of the discrete approximation s_δ in the time interval D of interest.

The qualitative behavior and the error norm of the reconstructed transient front function for Problem 3 is illustrated in Fig. 5.6 where the numerical solution for an average perturbation $\varepsilon = 0.005$ (full line) and the exact

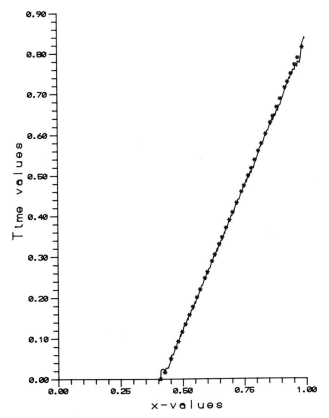

Fig. 5.6 Reconstructed front for Problem 3: $\varepsilon = 0.005$, $\delta = 0.04$, $h = k = 0.01$, $\|s - s_\delta\| = 0.0041$. Exact ($\ast \ast \ast$); computed (—).

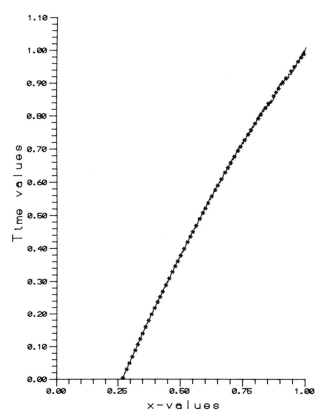

Fig. 5.7 Reconstructed front for Problem 4: $\varepsilon = 0.005$, $\delta = 0.04$, $h = k = 0.01$, $\|s - s_\delta\| = 0.0032$. Exact ($\ast\ast\ast$); computed (—).

solution are plotted. The computed front function (full line) and the corresponding discrete error norm given by (5.21), for Problem 4 and for the noise level $\varepsilon = 0.005$, are shown in Fig. 5.7.

Stefan Problem: Solid Phase. To emphasize the versatility of the method, in Problem 5 we consider a one-dimensional semi-infinite body, initially at zero temperature, exposed to a heat flux data function at the surface $x = 0$ given by

$$-u_x(0, t) = \begin{cases} [\pi(t - 0.2)]^{-1/2}, & t > 0.2, \\ 0, & 0 < t \le 0.2. \end{cases}$$

Hence, the exact temperature distribution has equation

$$u(x,t) = \begin{cases} \text{erfc}\{x/(2[t-0.2]^{1/2})\}, & x > 0, t > 0.2, \\ 0, & x > 0, 0 < t \le 0.2, \end{cases}$$

and in order to model the Stefan problem in the solid domain, we measure the data temperature function

$$u(1,t) = f(t) = \begin{cases} \text{erfc}\{1/(2[t-0.2]^{1/2})\}, & t > 0.2, \\ 0, & 0 < t \le 0.2, \end{cases}$$

and the data heat flux function

$$-u_x(1,t) = q(t)$$

$$= \begin{cases} [\pi(t-0.2)]^{-1/2}\exp\{-[4(t-0.2)]^{-1}\}, & t > 0.2, \\ 0, & 0 < t \le 0.2, \end{cases}$$

and attempt to determine the location of an isotherm $u(x,t) = \text{erfc}[x/(2t^{1/2})] = \text{constant}$, $0 \le x < 1$, $t \ge 0.2$. For instance, if we set $u(x,t) = 0.257899$, we find $x/[2(t^{1/2})] = 0.8$ and the exact location of the simulated "melting" front is given by

$$s(t) = \begin{cases} 1.6t^{1/2}, & t > 0.2, \\ 0, & 0 < t \le 0.2. \end{cases}$$

The computed front s_δ for this example, obtained by using the finite-difference scheme (5.10)—with the vertical and horizontal front-tracking procedures—now marching backward in space, is plotted in Fig. 5.8 (full line) together with the corresponding exact front location and the associated discrete l^2 error norm.

Notice that the noisy data functions are now given by

$$f_m(t_n) = f(t_n) + \varepsilon_{n,1}$$

and

$$q_m(t_n) = q(t_n) + \varepsilon_{n,2},$$

where t_n indicates a grid point in the data time interval $x = 1$ and $\varepsilon_{n,1}$ and $\varepsilon_{n,2}$ are Gaussian random variables of variance $\sigma^2 = \varepsilon$.

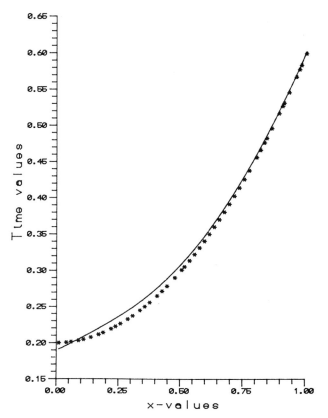

Fig. 5.8 Reconstructed front for Problem 5: $\varepsilon = 0.005$, $\delta = 0.04$, $h = k = 0.01$, $\|s - s_\delta\| = 0.0043$. Exact ($* * *$); computed (—).

Inverse Stefan Problem

For our last example, Problem 6, we utilize the inverse Stefan problem associated with Problem 3. We recall that in this case the exact location of the front is given by $s(t) = 2^{-1/2}t + 2^{1/2} - 1$, $0 < t \le T$, and for the formulation of the inverse Stefan problem, the two data functions are described by

$$u(s(t), t) = f(t) = 0, \qquad 0 < t \le T,$$

and

$$-u_x(s(t), t) = ds/dt(t) = q(t) = 2^{-1/2}, \qquad 0 < t \le T,$$

with initial temperature distribution given by

$$u(x, 0) = \exp[1 - 2^{-1/2} - 2^{-1/2}x] - 1, \qquad 0 \le x \le s(0).$$

In this problem, the task consists of recovering the temperature and the heat flux functions

$$u(0, t) = F(t) = \exp\left[1 - 2^{-1/2} + t/2\right] - 1, \qquad 0 < t < 1,$$

and

$$-u_x(0, t) = Q(t) = 2^{-1/2} \exp\left[1 - 2^{-1/2} + t/2\right], \qquad 0 < t < 1,$$

at the fixed liquid boundary $x = 0$, marching backward in space.

In this case, the vertical and horizontal search procedures for the numerical tracking of the melting front are no longer necessary. However, instead of starting at the front location $s(t)$, we first introduce the transformation $r = x/s(t)$, $t > 0$. In this manner, the parabolic problem in the new variables

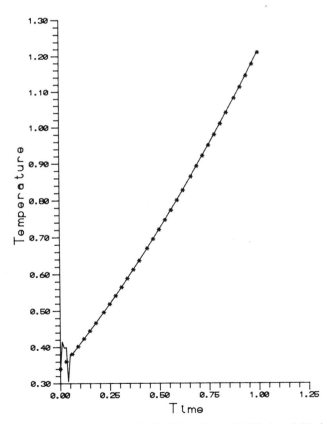

Fig. 5.9 Reconstructed temperature in Problem 6: $\varepsilon = 0.005$, $\delta = 0.04$, $h = k = 0.01$, $\|F - J_\delta F_m\| = 0.0038$. Exact ($* * *$); computed (—).

for the inverse Stefan problem becomes

$$u_{rr} = -rsu_r \, ds/dt + s^2 u_t, \qquad\qquad 0 < t < T, 0 < r < 1, s = s(t),$$
$$u(1,t) = f(t) = 0, \qquad\qquad 0 < t \le T,$$
$$-u_r(1,t) = q(t) = 2^{-1/2}(2^{-1/2}t + 2^{1/2} - 1), \qquad 0 < t \le T,$$
$$u(r,0) = \exp\left[1 - 2^{-1/2} - 2^{-1/2}r(2^{1/2} - 1)\right], \quad 0 \le r \le 1.$$

The finite-difference scheme consistent with the new partial differential equation is updated as follows:

$$W_i^n = \left[1 + (i + 1)h2^{-1/2}(nk2^{-1/2} + 2^{1/2} + 1)\right]W_{i+1}^n$$
$$- (nk2^{-1/2} + 2^{1/2} - 1)^2 \frac{h}{2k}\left(V_{i+1}^{n+1} - V_{i+1}^{n-1}\right),$$
$$V_i^n = V_{i+1}^n - hW_i^n, \qquad n = 1, 2, \ldots, M - 1, \qquad i = N - 1, \ldots, 1, 0.$$

Figures 5.9 and 5.10 illustrate the qualitative behavior and the discrete error norms of the computed temperature and heat flux grid solutions $\{V_0^n\}$

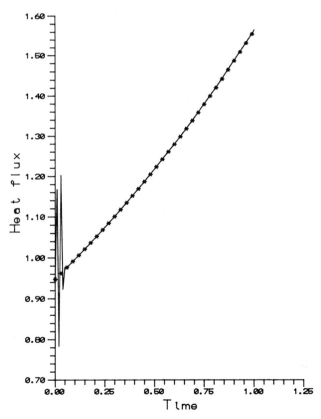

Fig. 5.10   Reconstructed heat flux in Problem 6: $\varepsilon = 0.005$, $\delta = 0.04$, $h = k = 0.01$, $\|Q - J_\delta Q_m\| = 0.0041$. Exact ($* * *$); computed (—).

and $\{W_0^n\}$, respectively, at the boundary location $x = 0$ (full line) for a noise level $\varepsilon = 0.005$. The exact temperature and heat flux solutions are also plotted for graphical comparison and we notice that the initial oscillations in the approximate solutions—notably for the reconstructed heat flux—disappear almost immediately as the method recovers very quickly.

In all cases, the numerical experiments corroborate the theoretical stability properties of the algorithm; it effectively restores stability with respect to perturbations in the data, which is essential, in particular, to justify the introduction of the inverse problem approach for the numerical solution of the Stefan problem, a well-posed problem.

5.6 IDENTIFICATION OF THE INITIAL TEMPERATURE DISTRIBUTION

We claim that the usual data for the one-dimensional IHCP discussed in Chap. 3 uniquely define the initial condition for the direct problem in the finite slab case. Moreover, it is also possible to analyze a suitable model problem for the semi-infinite conductor that permits us to solve the IHCP numerically and—at the same time—to identify the initial temperature distribution. These tasks are achieved by slightly modifying the usual space marching finite-difference scheme with the addition of an extrapolation "at the boundary" procedure.

5.7 SEMI-INFINITE BODY

We consider a one-dimensional semi-infinite conductor in which the temperature history on the active boundary surface $x = 0$ is desired and unknown, the initial temperature distribution is also desired and unknown in the space interval $0 \leq x \leq 1$, and the temperature history at the interior point $x = 1$ is approximately measurable together with the initial temperature distribution for $x > 1$.

Physically, this corresponds to the situation where there is no information whatsoever regarding a region—presumed inaccessible—surrounding the active boundary of the conductor.

Mathematically, assuming linear heat conduction with constant coefficients, the new inverse problem can be described as follows: the temperature solution $w(x, t)$ satisfies

$$w_t(x, t) = w_{xx}(x, t), \qquad t > 0, \, x > 0, \qquad (5.22a)$$

$$w(1, t) = \alpha(t), \qquad\qquad t > 0, \qquad (5.22b)$$

with corresponding approximate data function $\alpha_m(t)$,

$$w(x, 0) = \beta(x), \qquad t > 0, \tag{5.22c}$$

unknown for $0 \le x \le 1$, with corresponding approximate data function $\beta_m(x)$, $x > 1$,

$$w(0, t) = f(t), \qquad t > 0, \tag{5.22d}$$

the desired but unknown temperature function,

$$w(x, t) \to 0 \quad \text{as} \quad x \to \infty, t > 0. \tag{5.22e}$$

In order to be able to identify $f(t)$, $t > 0$, and $\beta(x)$, $0 \le x \le 1$, we consider first the intermediate direct problem

$$z_t(x, t) = z_{xx}(x, t), \qquad t > 0, x > 0, \tag{5.23a}$$

$$z(0, t) = 0, \qquad t > 0, \tag{5.23b}$$

$$z(x, 0) = \begin{cases} 0, & 0 \le x \le 1, \\ \beta_m(x), & x > 1, \end{cases} \tag{5.23c}$$

$$z(x, t) \to 0 \quad \text{as} \quad x \to \infty, t > 0. \tag{5.23d}$$

By superposition, the function $u(x, t)$—of the new inverse problem obtained by subtracting the solution of problem (5.23) from the solution of problem (5.22)—satisfies

$$u_t(x, t) = u_{xx}(x, t), \qquad\qquad t > 0, x > 0, \tag{5.24a}$$

$$u(1, t) = F(t) = \alpha(t) - z(1, t), \qquad t > 0, \tag{5.24b}$$

with corresponding approximate data function $F_m(t) = \alpha_m(t) - z(1, t)$,

$$u(x, 0) = \begin{cases} \beta(x), & 0 \le x \le 1, \text{unknown}, \\ 0, & x > 1, \end{cases} \tag{5.24c}$$

$$u(0, t) = f(t), \qquad t > 0, \text{unknown}, \tag{5.24d}$$

$$u(x, t) \to 0 \quad \text{as} \quad x \to \infty, t > 0. \tag{5.24e}$$

In what follows, system (5.24) will always be considered our prototype model, that is, we shall assume that problem (5.23) has been solved and the original data for the inverse new problem (5.22) have been updated accordingly, so that the inverse problem (5.24) is now feasible.

Remarks

1. Notice that the interior history temperature function $u(1, t) = F(t)$ in problem (5.24) can be thought of as "induced" by the initial data function $\beta(x)$ and the boundary data function $f(t)$ of the associated direct problem, and, as such, it has all the necessary data information for the solution of the inverse problem.

2. The prototype system (5.24) is somewhat intermediate between the classical backward-in-time heat conduction problem and the inverse heat conduction problem. We recall that for the backward-in-time problem, $u(x, T) = g(x)$, $T > 0$, $0 \leq x \leq 1$, $u(0, t) = f(t)$, $0 \leq t \leq T$, and $u(1, t) = F(t)$, $0 \leq t \leq T$, are given and the initial temperature distribution $u(x, 0) = \beta(x)$, $0 \leq x \leq 1$, is sought. For the IHCP instead, $u(x, 0) = \beta(x)$, $0 \leq x \leq 1$, and $u(1, t) = F(t)$, $t > 0$, are given and the boundary condition $u(0, t) = f(t)$, $t > 0$, needs to be identified.

Uniqueness. Before attempting to solve problem (5.24), it is important to show that the solution—assumed to exist on physical grounds—is unique.

To that effect, suppose that problem (5.24) admits two different solutions $u_1(x, t)$ and $u_2(x, t)$, analytic functions of their arguments x and t. Then the difference function $v = u_1 - u_2$ is such that

$$v_t(x, t) = v_{xx}(x, t), \qquad t > 0, x > 0,$$

$$v(0, t) \neq 0, \qquad t > 0,$$

$$v(x, 0) = 0, \qquad x \geq 1,$$

$$v(x, 0) \neq 0, \qquad 0 \leq x < 1,$$

$$v(1, t) = 0, \qquad t > 0.$$

Consequently, in the region $x \geq 1$, $t \geq 0$, $v(x, t) = 0$ and by the uniqueness of the analytic continuation procedure, $v(x, t) = 0$ everywhere in the quarter plane $x \geq 0$, $t \geq 0$. ∎

5.8 FINITE SLAB SYMMETRY

In the case of a finite-length conductor, the corresponding new inverse problem can be described as follows: the temperature solution $u(x, t)$ satisfies

$$u_t(x, t) = u_{xx}(x, t), \qquad 0 < x < 1, t > 0, \tag{5.25a}$$

$$u(1, t) = F(t), \qquad t > 0, \tag{5.25b}$$

with corresponding approximate data function $F_m(t)$,

$$-u_x(1,t) = Q(t), \qquad t > 0, \tag{5.25c}$$

with corresponding approximate data function $Q_m(t)$,

$$u(0,t) = f(t), \qquad t > 0, \text{ unknown}, \tag{5.25d}$$

$$-u_x(0,t) = q(t), \qquad t > 0, \text{ unknown}, \tag{5.25e}$$

$$u(x,0) = \beta(x), \qquad 0 \le x \le 1, \text{ unknown}. \tag{5.25f}$$

Remark. System (5.25), when the initial temperature distribution $\beta(x)$—in (5.25f)—is known, constitutes the usual statement of the IHCP for the one-dimensional finite slab geometry.

Uniqueness. Once again, we need to show that the system (5.25) has a unique solution.

Suppose that problem (5.25) admits two different solutions $u_1(x,t)$ and $u_2(x,t)$. Then the function $v = u_1 - u_2$ satisfies

$$v_t(x,t) = v_{xx}(x,t), \qquad 0 < x < 1, t > 0,$$

$$v(1,t) = 0, \qquad t > 0,$$

$$-v_x(1,t) = 0, \qquad t > 0,$$

$$v(0,t) = \Delta f(t) \ne 0, \qquad t > 0,$$

$$-v_x(x,t) = \Delta q(t) \ne 0, \qquad t > 0,$$

$$v(x,0) = \Delta\beta(x) \ne 0, \qquad 0 \le x \le 1.$$

If we imbed the slab domain $[0,1] \times [0,t)$ symmetrically into the slab domain $[0,2] \times [0,t)$, the relationship among the "boundary" data functions $v(0,t) = \Delta f$, $\tilde{v}(2,t) = \Delta \tilde{f}$, $-v_x(0,t) = \Delta q$, $-\tilde{v}_x(2,t) = \Delta\tilde{q}$, $v(x,0) = \Delta\beta$, $0 \le x \le 1$, and $\tilde{v}(x,0) = \Delta\tilde{\beta}$, $1 \le x \le 2$, is uniquely determined by the conditions that the function v must satisfy at $x = 1$ for all $t > 0$.

Given that we do not consider interior heat sources, the condition $v(1,t) = 0$, $t \ge 0$, can be achieved if and only if $\Delta f = -\Delta\tilde{f}$, $\Delta q = -\Delta\tilde{q}$, and $\Delta\beta = -\Delta\tilde{\beta}$. On the other hand, the condition $-v_x(1,t) = 0$, $t \ge 0$, can be obtained if and only if $\Delta f = \Delta\tilde{f}$, $\Delta q = \Delta\tilde{q}$, and $\Delta\beta = \Delta\tilde{\beta}$. Thus, it follows that

$$\Delta f = \Delta q = \Delta\beta \equiv 0$$

and this implies that the inverse problem (5.25) has a unique solution. ∎

5.9 STABILIZED PROBLEMS

Mollifying system (5.25), we obtain the following associated problem: attempt to find $J_\delta f_m(t) = J_\delta u(0, t)$, $J_\delta q_m(t) = -J_\delta u_x(0, t)$, and $J_\delta u(x, 0) = \beta(x)$ at the points of interest and for some radius $\delta > 0$, given that $J_\delta u(x, t)$ satisfies

$$
\begin{aligned}
(J_\delta u)_t &= (J_\delta u)_{xx}, & t > 0, x > 0, \\
J_\delta u(1, t) &= J_\delta F_m(t), & t > 0, \\
-J_\delta u_x(1, t) &= J_\delta Q_m(t), & t > 0, \\
J_\delta u(0, t) &= J_\delta f_m(t), & t > 0, \text{ unknown}, \\
-J_\delta u_x(0, t) &= J_\delta q_m(t), & t > 0, \text{ unknown}, \\
J_\delta u(x, 0) &= J_\delta \beta(x), & 0 \le x \le 1, \text{ unknown}.
\end{aligned}
\tag{5.26}
$$

The computational details and the necessary extrapolation procedure needed to numerically identify the initial temperature distribution $\beta(x)$ are presented next.

Numerical Procedure

With $v = J_\delta u$ and $z = -\partial v / \partial x$, system (5.26) is equivalent to

$$
\begin{aligned}
\frac{\partial v}{\partial t} &= -\frac{\partial z}{\partial x}, & t > 0, 0 < x < 1, \\
z &= -\frac{\partial v}{\partial x}, & t > 0, 0 < x < 1, \\
v(1, t) &= J_\delta F_m(t), & t > 0, \\
z(1, t) &= J_\delta Q_m(t), & t > 0, \\
v(0, t) &= J_\delta f_m(t), & t > 0, \text{ unknown}, \\
z(0, t) &= J_\delta q_m(t), & t > 0, \text{ unknown}, \\
v(x, 0) &= J_\delta \beta(x), & 0 \le x \le 1, \text{ unknown}.
\end{aligned}
\tag{5.27}
$$

Without loss of generality, we seek to reconstruct the unknown mollified boundary temperature function $J_\delta f_m$ and the mollified boundary heat flux function $J_\delta q_m$ in the unit interval $I_t = [0, 1]$ of the time axis $x = 0$ and the mollified initial temperature distribution $J_\delta \beta$ (x in the unit interval $I_x = [0, 1]$ of the space axis $t = 0$). Consider a uniform grid in the (x, t) space:

$$
\begin{aligned}
\{(x_i &= ih, t_n = nk), i = 0, 1, \ldots, N, Nh = 1; \\
n &= 0, 1, \ldots, M, Mk = L = 1 - k + k/h\}.
\end{aligned}
$$

Let the grid functions V and W be defined by

$$V_i^n = v(x_i, t_n), \qquad W_i^n = z(x_i, t_n), \qquad 0 \le i \le N, 0 \le n \le M.$$

Notice that

$$V_N^n = J_\delta F_m(t_n) \qquad \text{and} \qquad W_N^n = J_\delta Q_m(t_n), \qquad 0 \le n \le M.$$

We approximate the partial differential equation in system (5.27) with the consistent finite-difference scheme

$$W_{i-1}^n = W_i^n - \frac{h}{2k}(V_i^{n+1} - V_i^{n-1}),$$

$$V_{i-1}^n = V_i^n - hW_{i-1}^n, \qquad\qquad (5.28)$$

$$i = N, N-1, \ldots, 1, n = M-1, M-2, \ldots, 1,$$

obtaining a local truncation error that behaves as $O(h^2 + k^2)$ for the discrete mollified temperature and $O(h + k^2)$ for the discrete mollified heat flux as $h, k \to 0$.

We assume that the data functions F_m and Q_m are discrete functions measured at equally spaced sample points $t_j = jk$, $j = 0, 1, \ldots, M$, in $I_L = [0, L]$. We recall that in order to compute the discrete mollified functions $J_\delta F_m(t_j)$ and $J_\delta Q_m(t_j)$, we need to extend the data functions in such a way that F_m and Q_m decay smoothly to 0 in $I_{L, \delta_{max}} = [-\delta_{max}, L + \delta_{max}]$, $k \le \delta_{max} \le 0.1$, and both are 0 in $R - I_{L, \delta_{max}}$. In what follows, we consider the extended discrete data functions F_m and Q_m defined at equally spaced sample points on any time data interval of interest.

The selection of the radius of mollification is implemented as usual, solving the discrete equations

$$\|J_\delta F_m - F_m\|_{I_{L, \delta_{max}}} = \varepsilon$$

and

$$\|J_\delta Q_m - Q_m\|_{I_{L, \delta_{max}}} = \varepsilon,$$

using, for example, the bisection method.

The space marching technique discussed here requires the replacement of the discrete temperature and heat flux data functions F_m and Q_m at $x = 1$ by their filtered—and suitably extended—discrete mollified versions $J_\delta F_m$ and $J_\delta Q_m$, respectively. In most cases, the "compatibility" condition $\beta(1) = J_\delta F_m(0)$ will be lost in the process, even if the initial temperature function $\beta(x)$ is known and/or replaced by $J_\delta \beta(x)$ after mollifying the initial condition function in the x direction. As we march backward in space, the initial discrepancy might propagate generating a "boundary layer" effect that influences the initial values of the computed solutions $J_\delta f_m$ and $J_\delta q_m$ on the time interval $I_t = [0, 1]$, at the active interface $x = 0$.

However, it is possible to attempt to control this situation in a practical manner by replacing—after each step in space—the unknown mollified initial temperature $J_\delta \beta(ih) = V_i^0$, $i = N, N - 1, \ldots, 0$, by the linear extrapolation value obtained from the already estimated quantities V_i^1 and V_i^2, introducing an extra local error of order $O(k)$ in the sequel. The unknown approximate mollified initial temperature is calculated directly using the formula

$$V_i^0 = 2V_i^1 - V_i^2, \qquad i = N, N - 1, \ldots, 0.$$

Once the radii of mollification δ_F and δ_Q associated with the data functions F_m and Q_m, respectively, and the discrete filtered data functions $J_\delta F_m(t_n) = V_N^n$ and $J_\delta Q_m(t_n) = W_N^n$, $0 \le n \le M$, are determined, we apply the finite-difference algorithm (5.28)—modified "at the boundary" as explained previously—marching backward in the x direction with $\delta = \max(\delta_F, \delta_Q)$.

Remark. For the semi-infinite conductor, the numerical scheme remains exactly the same if the temperature *and* the heat flux histories are measured at the interior point $x = 1$. The implementation is quite similar with obvious changes if, instead, two temperature histories are measured at two different interior point locations $x = 1$ and $x = 1 - h$.

5.10 NUMERICAL RESULTS

In order to test the accuracy and stability properties of the method, in Problem 7 the approximate reconstruction of an initial temperature distribution function $u(x, 0) = \beta(x)$, $0 \le x \le 1$, is investigated for a one-dimensional finite slab exposed to a temperature data function at the free surface $x = 1$ given by $u(1, t) = F(t) = 1 + e^{-t}(\cos 1 + \sin 1)$, $t > 0$, and a heat flux data function $-u_x(1, t) = Q(t) = e^{-t}(\sin 1 - \cos 1)$, $t > 0$.

The exact temperature solution for this problem is

$$u(x, t) = 1 + e^{-t}(\cos x + \sin x), 0 \le x \le 1, t > 0,$$

and the initial temperature distribution to be approximately reconstructed has equation $u(x, 0) = \beta(x) = 1 + \cos x + \sin x$. The temperature and heat flux history functions at the $x = 0$ boundary are also unknowns for this inverse problem and the exact solutions are $u(x, t) = f(t) = 1 + e^{-t}$ and $-u_x = q(t) = e^{-t}(\sin x - \cos x)$.

In Problem 8 we attempt to approximately reconstruct the initial temperature function $\beta(x)$ for a finite slab with data functions $u(1, t) = F(t) = 0$, $t > 0$, and $-u_x(1, t) = Q(t) = \pi e^{-\pi^2 t}$, $t > 0$.

The unique temperature solution is given by $u(x, t) = e^{-\pi^2 t} \sin \pi x$. Thus, the exact unknown functions for the inverse problem are $u(x, 0) = \beta(x) =$

$\sin \pi x$, $0 \le x \le 1$, $u(0, t) = f(t) = 0$, $t > 0$, and $-u_x(0, t) = q(t) = -\pi e^{-\pi^2 t}$, $t > 0$.

The noisy data are obtained by adding a random error to the exact data at every grid point t_n in the interval $I_{L, \delta_{\max}}$:

$$F_m(t_n) = F(t_n) + \varepsilon_{n,1},$$

$$Q_m(t_n) = Q(t_n) + \varepsilon_{n,2},$$

where $\varepsilon_{n,1}$ and $\varepsilon_{n,2}$ are Gaussian variables of variance $\sigma^2 = \varepsilon^2$.

In all cases, we use $h = \Delta x = 0.01$ and $k = \Delta t = 0.01$.

If the discretized computed initial temperature function component is denoted by $\beta^n_{\delta, \varepsilon}$ and the true component is $\beta^n = \beta(t_n)$, we use the sample

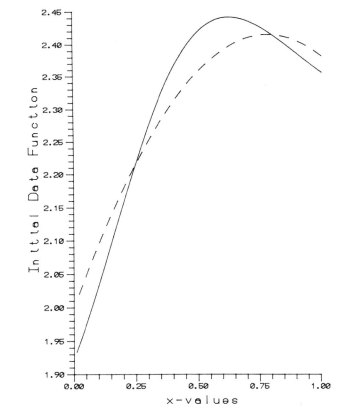

Fig. 5.11 Reconstructed initial temperature in Problem 7: $\varepsilon = 0.005$, $\delta = 0.04$, $h = k = 0.01$, $\|\beta_{\delta, \varepsilon} - \beta\| = 0.0421$. Exact (- - -); computed (—).

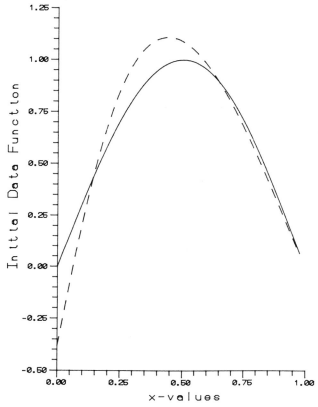

Fig. 5.12 Reconstructed initial temperature in Problem 8: $\varepsilon = 0.005$, $\delta = 0.04$, $h = k = 0.01$, $\|\beta_{\delta, \varepsilon} - \beta\| = 0.0931$. Computed (- - -); exact (—).

root mean square norm to measure the error in the discretized interval $I_x = [0, 1]$. The solution error is then given by

$$\|\beta_{\delta, \varepsilon} - \beta\|_{I_x} = \left[\frac{1}{M - N} \sum_{n = 1}^{M - N} (\beta_{\delta, \varepsilon}^n - \beta^n)^2 \right]^{1/2}.$$

The solution error for the reconstructed temperature and heat flux functions in the discretized interval $I_t = [0, 1]$ is evaluated similarly.

The qualitative behavior of the computed initial temperature functions for Problems 7 and 8 are illustrated in Figs. 5.11 and 5.12, respectively, for an average perturbation—level of noise—$\varepsilon = 0.005$ and radius of mollification

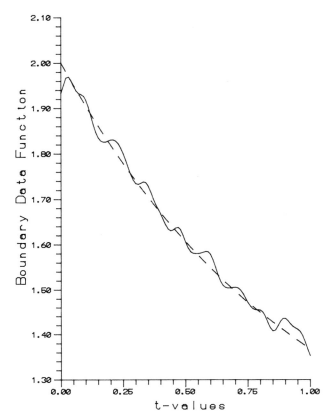

Fig. 5.13 Reconstructed boundary temperature in Problem 7: $\varepsilon = 0.005$, $\delta = 0.04$, $h = k = 0.01$, $\|f_{\delta,\varepsilon} - f\| = 0.0245$. Exact (- - -); computed (—).

$\delta = 0.04$. For completeness, in Fig. 5.13 we also show the reconstructed boundary temperature for Problem 7.

Remarks

1. The space marching computational procedure for the solution of the IHCP—modified at each step in space by performing a linear extrapolation "at the boundary"—can be implemented to evaluate—in a stable manner—the unknown initial temperature.
2. In certain situations, the modified algorithm might constitute a viable alternative for the solution of the backward-in-time inverse heat conduction problem.

5.11   EXERCISES

5.1. Consider system (5.1) with the condition (5.1f) replaced by

$$-u_x(0, t) = q(t) = \alpha(t)E(u(0, t)) - G(t), \qquad t > 0,$$

where $\alpha(t)$ is the transient heat transfer coefficient at the surface $x = 0$. Define the approximate transfer coefficient $\alpha_{\delta,\varepsilon}(t)$ by

$$\alpha_{\delta,\varepsilon}(t) = \frac{J_\delta q_m(t) + G(t)}{E(J_\delta f_m(t))}, \qquad t > 0. \tag{5.29}$$

Under the conditions of Theorem 3.4, if $E(u(0, t)) \geq \lambda > 0$ and E has a Lipschitz constant L with respect to the L^2 norm, show that in any time interval of interest, $D = [0, T]$,

$$\|\alpha - \alpha_{\delta,\varepsilon}\|_{2, D} \leq C\left\{\tfrac{1}{2}M\delta + 2\varepsilon \exp\left[(2\delta)^{-2/3}\right]\right\}, \tag{5.30}$$

where $C = C(\lambda, L, \|J_\delta q_m\|_{\infty, D}, \|G\|_{\infty, D}, \|E\|_{\infty, D})$.

5.2. Prove that it is possible to choose $\delta = \delta(\varepsilon)$ such that in (5.30), $\|\alpha - \alpha_{\delta,\varepsilon}\|_{2, D} \to 0$ as $\varepsilon \to 0$.

5.3. Implement the mollification method to numerically solve the IHCP

$$u_t = \tfrac{1}{2}e^t u u_{xx}, \qquad\qquad 0 < x < 1, 0 < t < 1,$$

$$-u_x(1, t) = Q(t) = -3e^t, \qquad 0 < t \leq 1,$$

$$u(1, t) = F(t) = 4e^t, \qquad 0 < t \leq 1,$$

$$u(x, 0) = x^2 + x + 2, \qquad 0 \leq x \leq 1.$$

Set $h = \Delta x = k = \Delta t = 0.01$ and use discrete random perturbations to obtain F_m and Q_m with $\varepsilon = 10^{-3}$ and 5×10^{-3}. In each case, find the discrete mollified temperature and heat flux functions $J_\delta f_m$ and $J_\delta q_m$, respectively, at the active surface $x = 0$.

5.4. If the exact heat flux function at $x = 0$ in Exercise 5.3 is given by $q(t) = -u_x(0, t) = \alpha(t)u(0, t) + G(t)$ and

$$G(t) = \begin{cases} 0, & 0 \leq t < 1/2, \\ e^t, & 1/2 \leq t \leq 1, \end{cases}$$

compute the discrete version of $\alpha_{\delta, \varepsilon}$ from (5.29). Compare your results with the exact values—at the grid points of the unit time interval—given by

$$\alpha(t) = \begin{cases} 1/2, & 0 \le t < 1/2, \\ 0, & 1/2 \le t \le 1. \end{cases}$$

5.12 REFERENCES AND COMMENTS

The following references and comments serve to expand the basic material covered in the corresponding sections.

5.1. The existence and uniqueness of a strictly increasing solution for the semi-infinite body version of the problem of determining the interface temperature between a gas and a solid under nonlinear monotone Lipschitz radiation laws was first considered by

W. R. Mann and F. Wolf, Heat transfer between solids and gases under nonlinear boundary conditions, *Quart. Appl. Math.* **9** (1951), 163–184.

The Lipschitz condition on the nonlinear heat transfer law was removed in the subsequent work of

J. H. Roberts and W. R. Mann, On a certain nonlinear integral equation of Volterra type, *Pacific J. Math.* **1** (1951), 431–445.

A constructive proof for the existence and uniqueness of the interface temperature in the presence of a positive integrable transient source by the method of lower and upper solutions can be found in

J. B. Keller and W. E. Olmstead, Temperature of a nonlinearly radiating semi-infinite solid, *Quart. Appl. Math.* **30** (1972), 559–566.

The numerical solution of the Volterra integral equation characterizing the active surface temperature history was successfully implemented, using the method of successive approximations, by

P. L. Chambré, Nonlinear heat transfer problem, *J. Appl. Phys.* **10** (1959), 1683–1688.

More recently, combined Abel inversion formulas and *B*-spline approximations together with product integration quadrature techniques appeared in

C. W. Groetsch, Convergence of a numerical algorithm for a nonlinear heat transfer problem, *Z. Angew. Math. Mech.* **65** (1985), 645.

C. W. Groetsch, A simple numerical model for nonlinear warming on a slab, in *Proc. of the International Symposium on Computational Mathematics*, T. Yamamoto (ed.), North-Holland, Amsterdam, 1991, pp. 149–156.

More general problems of the same kind, combining the effects of convection and radiation at the interface, are described in

A. Friedman, Generalized heat transfer between solids and gases, under nonlinear boundary conditions, *J. Math. Mech.* **8** (1959), 161–183.

V. Saljnikov and S. Petrovic, Heating problem of a horizontal semi-infinite solid by natural convection, *Z. Angew. Math. Mech.* **68** (1988), 58–59.

5.2. Here we follow in part the work of

D. A. Murio, Numerical identification of interface source functions in heat transfer problems under nonlinear boundary conditions, *Comput. Math. Appl.* **24** (1992), 65–76.

5.3. The IHCP approach for the numerical solution of Stefan's problem was initially implemented, using space marching moving finite elements, by

M. A. Katz and B. Rubinsky, An inverse finite element technique to determine the change of phase interface location in one-dimensional melting problems, *Numer. Heat Transfer* **7** (1984), 269–283.

Space marching finite-difference techniques were utilized in

M. Raynaud, Comparison of space marching finite difference technique and function minimization technique for the estimation of the front location in nonlinear melting problems, *Fifth Symp. on Control of Distributed Parameter Systems*, A. El Jai and M. Amoroux (eds.), Université de Perpignan, France, 1989, pp. 215–220.

where a nonlinear version of the problem is investigated and some numerical experiments are presented.

For the inverse Stefan problem, a combined numerical procedure based on the integral method and the future temperatures algorithm was investigated by

N. Zabaras, S. Mukherjee, and O. Richmond, An analysis of inverse heat transfer problems with phase changes using an integral method, *ASME J. Heat Transfer* **110** (1988), 554–561.

The numerical solution of the inverse Stefan problem was also implemented using nonlinear approximation techniques of optimal control in a series of notes:

P. Jochum, The numerical solution of the inverse Stefan problem, *Numer. Math.* **34** (1980), 411–429.

P. Jochum, The inverse Stefan problem as a problem of nonlinear approximation theory, *J. Approx. Theory* **30** (1980), 81–89.

5.4. For the one-dimensional Stefan problem, there are numerous results available concerning the existence, uniqueness, and stability of the solution.

For a comprehensive review, see

L. I. Rubinstein, *The Stefan Problem*, American Mathematical Society, Providence, 1971.

A detailed description of several direct methods for the approximate solution of the Stefan problem and some selected algorithms for the numerical solution of the inverse Stefan problem can be found in the classical book of

J. Crank, *Free and Moving Boundary Problems*, Clarendon, Oxford, 1984.

5.5. In this section we follow in part the article

D. A. Murio, Solution of inverse heat conduction problems with phase changes by the mollification method, *Comput. Math. Appl.* **24** (1992), 45–57.

5.6–5.8. For specific information about the numerical solution of the back-ward-in-time inverse heat conduction problem, see the interesting article

B. L. Buzbee and A. Carasso, On the numerical computation of parabolic problems for preceding times, *Math. Comp.* **27** (1973), 237–266.

A discussion of more general backward-in-time parabolic problems can be found in

K. Miller, Efficient numerical methods for backward solution of parabolic problems with variable coefficients, *Improperly Posed Boundary Value Problems*, A. Carasso and A. P. Stone (eds.), Pitman, London, 1975, pp. 54–64.

Basic ideas about analytic continuation are discussed—at an elementary but rigorous level—in the well-known textbook

R. V. Churchill and J. V. Brown, *Complex Variables and Applications*, McGraw-Hill, New York, 4th ed., 1984.

5.9–5.10. In these sections we follow in part the original presentation given in

D. A. Murio and D. Hinestroza, On the space marching solution of the inverse heat conduction problem and the identification of the initial temperature distribution, *Comput. Math. Appl.*, **25** (1993), pp. 55–63.

6

APPLICATIONS OF STABLE NUMERICAL DIFFERENTIATION PROCEDURES

In this chapter we investigate several ill-posed identification problems in ordinary and partial differential equations—appearing in various important problems in science and engineering—that can be solved numerically by utilizing methods based on the stable computation of derivatives when the experimental information is obtained through measured data.

In all cases we use the mollification method—as introduced in Chap. 1—to develop completely automated algorithms for the approximate reconstruction of the unknown functions in a suitable compact subset of the domain where the solution is observed. The compact subset is automatically determined by the amount of noise in the data.

We begin by attempting to reconstruct the unknown forcing term in a system of ordinary differential equations—a problem related to perturbation theory and validation of physical structures—and then proceed with the numerical identification of the transmissivity coefficient in linear and nonlinear elliptic and parabolic partial differential equations, a quite general problem arising, for example, in porous media and related to ground water modeling and reservoir simulation. For clarity, we restrict our attention to the one-dimensional case in space. References to the extension of these techniques to two-dimensional problems are provided at the end of this chapter.

6.1 NUMERICAL IDENTIFICATION OF FORCING TERMS

We consider a system of ordinary differential equations of the form

$$\frac{dy}{dt}(t) = f(t, y(t)) + p(t) \tag{6.1}$$

together with the corresponding initial and/or boundary conditions (IC–BC), where

$$y(t) = \begin{pmatrix} y_1(t) \\ y_2(t) \\ \vdots \\ y_n(t) \end{pmatrix} = (y_1(t), y_2(t), \ldots, y_n(t))^T,$$

$$f(t, y(t)) = (f_1(t, y(t)), f_2(t, y(t)), \ldots, f_n(t, y(t)))^T,$$

$$p(t) = (p_1(t), p_2(t), \ldots, p_n(t))^T,$$

and

$$\frac{dy}{dt}(t) = \left(\frac{dy_1}{dt}(t), \frac{dy_2}{dt}(t), \ldots, \frac{dy_n}{dt}(t) \right)^T.$$

In general, f and p are known functions and we are required to find a solution function y that satisfies (6.1) and the (IC–BC). This constitutes the direct problem. However, it is possible to attempt to gain some knowledge about the structure of the system, for example the forcing term $p(t)$, from experimental information given by the approximate knowledge of the solution $y(t)$ at a discrete set of points in some interval of interest. This problem belongs to a general class of inverse problems, known as system identification problems, and, in particular, it is an ill-posed problem because small errors in the data function $y(t)$ might cause large errors in the computation of the derivative function $dy/dt(t)$ which is needed in order to estimate the forcing term function $p(t)$.

Our method begins by attempting to reconstruct a mollified version of the derivative function. Once this has been computed, the filtered data function is used to evaluate the function f and estimate the unknown forcing term. In the process we need to generalize the parameter selection criterion—introduced in Sec. 1.4 for the scalar case—to the vector case.

6.2 STABILIZED PROBLEM

In what follows we assume that the solution function $y(t)$ is measured in the interval $I = [0, 1]$. On the basis of this information, we discuss the problem of estimating the forcing term function $p(t)$ in some suitable compact set K, $K \subset I$.

If $C^0(I, \mathbb{R}^n)$ denotes the set of continuous vector functions over I with

$$\|y\|_{\infty, I} = \max_{1 \leq i \leq n} \|y_i\|_{\infty, I}, \qquad y_i \in C^0(I), \qquad i = 1, 2, \ldots, n,$$

where $C^0(I)$ represents the set of continuous scalar functions over I with

$$\|y_i\|_{\infty, I} = \max_{t \in I} |y_i(t)|,$$

we assume that $p(t) \in C^0(K, \mathbb{R}^n)$, $d^2 y(t)/dt^2 \in C^0(K, \mathbb{R}^n)$, and the function $f(t, y(t))$ has a Lipschitz constant L with respect to the norm $\|\cdot\|_\infty$. We also assume that instead of the function $y(t)$, we know some data function $y_m(t) \in C^0(I, \mathbb{R}^n)$ such that

$$\|y_m - y\|_{\infty, I} \leq \varepsilon. \tag{6.2}$$

In order to stabilize the differentiation problem, if $\delta > 0$ is smaller than the distance from K to the boundary of I, ∂I, we introduce the function

$$J_\delta y(t) = (\rho_\delta * y)(t) = \begin{pmatrix} \int_{-\infty}^{\infty} \rho_\delta(t - s) y_1(s)\, ds \\ \vdots \\ \int_{-\infty}^{\infty} \rho_\delta(t - s) y_n(s)\, ds \end{pmatrix} = \int_{-\infty}^{\infty} \rho_\delta(t - s) y(s)\, ds.$$

$J_\delta y(t)$ is a C^∞ function in $\mathbb{R}^n$ and for fixed $t \in K$ it has compact support in the interval I.

The following two lemmas are needed for our stability analysis. Their proofs, given in Sec. 1.2 for the scalar case, extend naturally to the vector function situation.

Lemma 6.1 (Consistency)   If $\|d^2 y/dt^2\|_{\infty, K} \leq M_2$, then

$$\left\| \frac{d}{dt} (\rho_\delta * y) - \frac{d}{dt} y \right\|_{\infty, K} \leq 3\delta M_2.$$

Lemma 6.2 (Stability)   If $y_m(t) \in C^0(I, \mathbb{R}^n)$ and $\|y_m - y\|_{\infty, K} \leq \varepsilon$, then

$$\left\| \frac{d}{dt} (\rho_\delta * y_m) - \frac{d}{dt} (\rho_\delta * y) \right\|_{\infty, K} \leq \frac{2\varepsilon}{\delta\sqrt{\pi}}.$$

We immediately have the following result.

Theorem 6.1 (Error Estimate)   Under the conditions of Lemmas 6.1 and 6.2,

$$\left\| \frac{d}{dt}(\rho_\delta * y_m) - \frac{d}{dt}y \right\|_{\infty, K} \le 3\delta M_2 + \frac{2\varepsilon}{\delta\sqrt{\pi}}. \tag{6.3}$$

We can now state our main theoretical result.

Theorem 6.2   Under the conditions of Lemmas 6.1 and 6.2, if $f(t, y)$ has a Lipschitz constant L, then

$$\| p_{\delta, \varepsilon} - p \|_{\infty, K} \le 3\delta M_2 + \varepsilon \left[\frac{2}{\delta\sqrt{\pi}} + L \right], \tag{6.4}$$

where

$$p_{\delta, \varepsilon} = \frac{d}{dt}(\rho_\delta * y_m)(t) - f(t, y_m(t)) \tag{6.5}$$

is the approximate reconstructed forcing term.

Proof. From (6.1) and (6.5), we have

$$p(t) - p_{\delta, \varepsilon}(t) = \frac{d}{dt}y(t) - \frac{d}{dt}(\rho_\delta * y_m)(t) + f(t, y_m(t)) - f(t, y(t)).$$

Taking norms, using the Lipschitz property of f and inequality (6.3), we obtain

$$\| p_{\delta, \varepsilon} - p \|_{\infty, K} \le 3\delta M_2 + \frac{2\varepsilon}{\delta\sqrt{\pi}} + L\| y_m - y \|_{\infty, K}.$$

By (6.2),

$$\| p_{\delta, \varepsilon} - p \|_{\infty, K} \le 3\delta M_2 + \varepsilon \left[\frac{2}{\delta\sqrt{\pi}} + L \right]. \quad \blacksquare$$

We observe that the right-hand side of inequality (6.4) cannot, in general, be minimized because M_2 is not known. Instead, we apply the procedure described previously in Sec. 1.4 to each component y_i, $1 \le i \le n$, and set $\delta = \min_{1 \le i \le n} \delta_i$, where δ_i is the unique solution of the equation

$$\| p_\delta * y_{m,i} - y_i \|_{\infty, I} = \varepsilon, \qquad 1 \le i \le n.$$

It is clear—from the monotonicity property (Lemma 1.3)—that $\tilde{\delta}$ is such that

$$\| J_{\tilde{\delta}} y_m - y_m \|_{2, I} \le \varepsilon. \tag{6.6}$$

Notice that the choice of $\delta = \tilde{\delta}$ automatically defines the compact subset $K = [3\tilde{\delta}, 1 - 3\tilde{\delta}]$ where we seek to reconstruct the unknown forcing term $p(t)$. We also notice that (6.6) justifies replacing $f(t, y_m(t))$ by $f(t, (\rho_\delta * y_m)(t))$, when attempting to reconstruct the approximate forcing term $\rho_{\delta, \varepsilon}$ using formula (6.5).

6.3   NUMERICAL RESULTS

The numerical method is implemented exactly as explained in Sec. 1.5 with the updates given previously. As usual, the discrete noisy data function is obtained by adding a random error to the exact data function, that is,

$$y_{m,i}(t_j) = y_i(t_j) + \varepsilon \theta_{ij},$$

where $t_j = j \Delta t$, $j = 0, 1, \ldots, N$, $N\Delta t = 1$, $i = 1, 2, \ldots, n$, and θ_{ij} is a Gaussian variable with values in $[-1, 1]$. If the discrete numerical approximation to the forcing term components $p_i(t_j)$ are denoted $P_i(t_j) = p_{\tilde{\delta}, \varepsilon, i}(t_j)$, $1 \le i \le n$, after computing $d/dt(\rho_{\tilde{\delta}} * y_{m,i})(t_j)$, we use the equation

$$P_i(t_j) = \frac{d}{dt}(\rho_{\tilde{\delta}} * y_{m,i})(t_j) - f_i\big(t_j, (\rho_{\tilde{\delta}} * y_{m,i})(t_j)\big)$$

on the sample points of the interval $K_{\tilde{\delta}} = [3\tilde{\delta}, 1 - 3\tilde{\delta}]$. The solution error is then given by

$$\| p_i - P_i \|_{2, K_{\tilde{\delta}}} = \left[\frac{1}{M} \sum_{j_{\min}}^{j_{\max}} |p_i(t_j) - P_i(t_j)|^2 \right]^{1/2}, \qquad i = 1, 2, \ldots, n,$$

where M indicates the number of sample points in $K_{\tilde{\delta}}$.

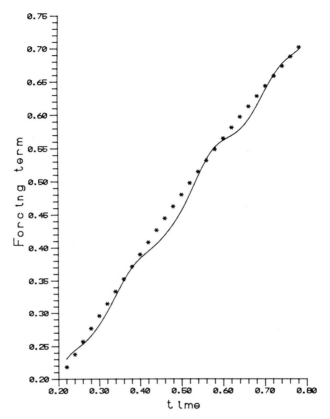

Fig. 6.1 Reconstructed forcing term in Example 1: $\varepsilon = 0.005$, $\delta = 0.06$, $\Delta t = k = 0.01$, $\|p - P\| = 0.00097$. Exact ($* * *$); computed (—).

EXAMPLE 1   As a first example we consider the scalar initial value problem $dy/dt = y + \sin t$, $y(0) = 1$. The exact solution is $y(t) = \frac{1}{2}[3e^t - \cos t - \sin t]$. Figure 6.1 shows the solution obtained with the mollification method and the exact forcing term function $p(t) = \sin t$. With $\varepsilon = 0.005$ and $\Delta t = 0.01$, it follows that $\tilde{\delta} = 0.06$ and $\|p_1 - P_1\|_{2, K_{\tilde{\delta}}} = 0.00097$.

EXAMPLE 2   The second example is an initial value problem given by the system $dy_1/dt = 5y_1 - 6y_2 + t + 3$, $dy_2/dt = 3y_1 - 4y_2 + t + 2$, $y_1(0) = 1$, $y_2(0) = 1$. The exact solution is $y_1 = e^{2t} + e^{-t} + t - 1$, $y_2(t) = \frac{1}{2}e^{2t} + e^{-t} + t - \frac{1}{2}$. Figures 6.2$a$ and b show the numerical solutions and the exact forcing functions $p_1(t) = t + 3$ and $p_2(t) = t + 2$, respectively, corresponding to $\varepsilon = 0.005$, $\Delta t = 0.01$, and $\tilde{\delta} = 0.06$. The associated error norms are given by $\|p_1 - P_1\|_{2, K_{\tilde{\delta}}} = 0.00101$ and $\|p_2 - P_2\|_{2, K_{\tilde{\delta}}} = 0.00099$.

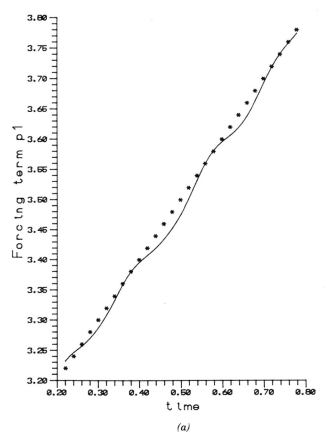

(a)

Fig. 6.2a Reconstructed forcing term p_1 in Example 2: $\varepsilon = 0.005$, $\delta = 0.06$, $\Delta t = k = 0.01$, $\|p_1 - P_1\| = 0.00101$. Exact ($* * *$); computed (—).

6.4   IDENTIFICATION OF THE TRANSMISSIVITY COEFFICIENT IN THE ONE-DIMENSIONAL ELLIPTIC EQUATION

The identification problem can be described as follows.

If the solution function $u(x)$ and the source function $f(x)$ are measured in the space interval $I = [0, 1]$, it is desired to identify the transmissivity coefficient $\alpha(x)$ such that

$$\begin{cases} \dfrac{d}{dx}\left(\alpha(x)\dfrac{du}{dx}(x)\right) = f(x), & 0 < x < 1, \\ \alpha(0)\dfrac{du}{dx}(0) = c, \end{cases} \tag{6.7}$$

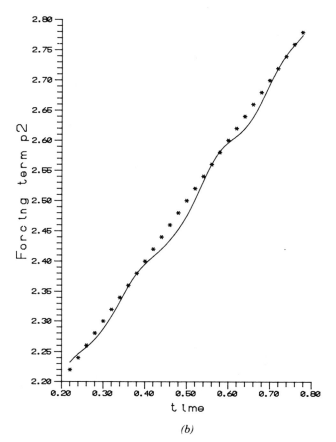

Fig. 6.2b Reconstructed forcing term p_2 in Example 2: $\varepsilon = 0.005$, $\delta = 0.06$, $\Delta t = k = 0.01$, $\|p_2 - P_2\| = 0.00099$. Exact ($* * *$); computed (—).

in the linear case, and

$$\begin{cases} \dfrac{d}{dx}\left(\alpha(u(x))\dfrac{du}{dx}(x)\right) = f(x), & 0 < x < 1, \\[4mm] \alpha(u(0))\dfrac{du}{dx}(0) = c, \end{cases} \tag{6.8}$$

in the nonlinear case.

We study these two problems under each of the following assumptions, valid over a suitable compact set $K \subset I$:

$$\inf_K\left\{\left\|\frac{d}{dx}u(x)\right\|\right\} > 0 \tag{A1}$$

or

$$\inf_{K} \left\{ \max \left(\left| \frac{du}{dx}(x) \right|, \frac{d^2u}{dx^2}(x) \right) \right\} > 0. \tag{A2}$$

Assumption (A2) eliminates the possibility that $du/dx(x)$ and $d^2u/dx^2(x)$ vanish simultaneously and also the vanishing of $du/dx(x)$ at more than one point in K. We also assume that the exact and measured data functions satisfy error bounds of the form

$$\|f - f_m\|_{\infty, I} \le \varepsilon, \tag{6.9}$$

$$\|u - u_m\|_{\infty, I} \le \varepsilon. \tag{6.10}$$

6.5 STABILITY ANALYSIS

Under assumption (A1), (6.7) and (6.8) can be integrated trivially, obtaining the unique solutions

$$\alpha(x) = \frac{\displaystyle\int_0^x f(s)\,ds + c}{\displaystyle\frac{d}{dx}u(x)} \tag{6.11}$$

and

$$\alpha(u(x)) = \frac{\displaystyle\int_0^x f(s)\,ds + c}{\displaystyle\frac{d}{dx}u(x)}, \tag{6.12}$$

respectively. Since both formulas are similar, we devote all our attention to the linear case.

The approximate transmissivity coefficients $\alpha_\delta(x)$ and $\alpha_{\delta, \varepsilon}(x)$ are respectively defined by

$$\alpha_\delta(x) = \frac{\displaystyle\int_0^x f(s)\,ds + c}{\displaystyle\frac{d}{dx}J_\delta u(x)} \tag{6.13}$$

and

$$\alpha_{\delta,\varepsilon}(x) = \frac{\int_0^x f_m(s)\,ds + c}{\dfrac{d}{dx} J_\delta u_m(x)}. \tag{6.14}$$

We now establish some preliminary results.

Lemma 6.3 (Consistency)   If

$$\left| \frac{d^2}{dx^2} u(x) \right| \leq M_2 \quad \text{and} \quad \left| \frac{d}{dx} u(x) \right| \geq \lambda > 0, \qquad x \in K_\delta \subset I,$$

then

$$\|\alpha - \alpha_\delta\|_{\infty, K_\delta} \leq C\delta, \tag{6.15}$$

for some constant

$$C = C\left(\lambda, M_2, \left\| \frac{d}{dx} u \right\|_{\infty, K_\delta}, \|f\|_{\infty, I}, |c| \right).$$

Proof. From (6.11) and (6.13), for $x \in K_\delta$, we have

$$|\alpha(x) - \alpha_\delta(x)| = \left| \left(\int_0^x f(s)\,ds + c \right) \left(\frac{1}{\dfrac{d}{dx} u(x)} - \frac{1}{\dfrac{d}{dx} J_\delta u(x)} \right) \right|$$

$$= \left| \frac{\left(\int_0^x f(s)\,ds + c \right) \left(\dfrac{d}{dx} J_\delta u(x) - \dfrac{d}{dx} u(x) \right)}{\dfrac{d}{dx} J_\delta u(x) \dfrac{d}{dx} u(x)} \right|$$

$$\leq \frac{1}{\lambda^2} (|c| + \|f\|_{\infty, I}) \left\| \frac{d}{dx} J_\delta u - \frac{d}{dx} u \right\|_{\infty, K_\delta}$$

$$\leq \frac{1}{\lambda^2} (|c| + \|f\|_{\infty, I}) 3\delta M_2$$

$$= C\delta$$

by Lemma 1.1.   ■

Lemma 6.4 (Stability) If u_m and f_m verify inequalities (6.9) and (6.10),

$$\left| \frac{d}{dx} u(x) \right| \geq \lambda > 0, \, x \in K_\delta \subset I, \quad \text{and} \quad \delta > \frac{2\varepsilon}{\lambda \sqrt{\pi}},$$

then

$$\| \alpha_\delta - \alpha_{\delta, \varepsilon} \|_{\infty, K_\delta} \leq \frac{C\varepsilon}{\lambda(\lambda \delta \sqrt{\pi} - 2\varepsilon)}(\lambda \delta + 1), \qquad (6.16)$$

where $C = \max\{2(|c| + \|f\|_{\infty, I}), \sqrt{\pi}\}$.

Proof. For $x \in K_\delta$, from (6.13) and (6.14), we have

$|\alpha_\delta(x) - \alpha_{\delta, \varepsilon}(x)|$

$$= \left| \left(\int_0^x f(s)\, ds + c \right) \left(\frac{1}{\frac{d}{dx} J_\delta u(x)} - \frac{1}{\frac{d}{dx} J_\delta u_m(x)} + \frac{\int_0^x (f(s) - f_m(s))\, ds}{\frac{d}{dx} J_\delta u_m(x)} \right) \right|$$

$$\leq \left| \frac{\int_0^x f(s)\, ds + c}{\frac{d}{dx} J_\delta u(x) \frac{d}{dx} J_\delta u_m(x)} \right| \left| \frac{d}{dx} J_\delta u_m(x) - \frac{d}{dx} J_\delta u(x) \right| + \frac{\varepsilon}{\left| \frac{d}{dx} J_\delta u_m(x) \right|}$$

$$\leq \frac{(|c| + \|f\|_{\infty, I})}{\lambda \left(\lambda - \frac{2\varepsilon}{\delta \sqrt{\pi}} \right)} \frac{2\varepsilon}{\delta \sqrt{\pi}} + \frac{\varepsilon}{\lambda - \frac{2\varepsilon}{\delta \sqrt{\pi}}}$$

$$\leq \frac{C\varepsilon}{\lambda(\lambda \delta \sqrt{\pi} - 2\varepsilon)}(\lambda \delta + 1),$$

using Lemma 1.2 and setting $C = \max\{2(|c| + \|f\|_{\infty, I}), \sqrt{\pi}\}$. ∎

We immediately have the following result.

Theorem 6.3 (Error Estimate) Under the conditions of Lemmas 6.3 and 6.4,

$$\| \alpha - \alpha_{\delta, \varepsilon} \|_{\infty, K_\delta} \leq C \left(\delta + \frac{\varepsilon(\lambda \delta + 1)}{\lambda(\lambda \delta \sqrt{\pi} - 2\varepsilon)} \right). \qquad (6.17)$$

Remarks

1. Notice that if we set $\delta = O(\sqrt{\varepsilon})$, for instance, we get uniform convergence, that is, $\| \alpha - \alpha_{\delta, \varepsilon} \|_{\infty, K_\delta} \to 0$ as $\varepsilon \to 0$.

2. Similar bounds for the approximate transmissivity coefficient are obtained in the nonlinear case.

Under assumption (A2), where u has a unique critical point x^* at an interior point of K_δ, for all x in K_δ, $x \neq x^*$, we have

$$\alpha(x) = \frac{\int_{x^*}^x f(s)\, ds}{\dfrac{d}{dx} u(x)}, \tag{6.18}$$

$$\alpha_\delta(x) = \frac{\int_{x^*}^x f(s)\, ds}{\dfrac{d}{dx} J_\delta u(x)}, \tag{6.19}$$

and

$$\alpha_{\delta,\varepsilon}(x) = \frac{\int_x^x f_m(s)\, ds}{\dfrac{d}{dx} J_\delta u_m(x)} \tag{6.20}$$

in the linear case.

Also in the linear case, at $x = x^*$, we have

$$\alpha(x^*) = \frac{f(x^*)}{\dfrac{d^2}{dx^2} u(x^*)}. \tag{6.21}$$

For the nonlinear case we proceed similarly.

We observe that under assumption (A2) no boundary conditions are needed. Moreover, we can select an open interval $V(x^*) = (x^* - \gamma, x^* + \gamma)$ such that $|d/dx\, u(x)| \geq \lambda_{x^*} > 0$ for all $x \in K_\delta - V(x^*)$. In the next section we will explain how to approximate the value of α at any point in $V(x^*)$ and avoid the computation of the second derivative in (6.21).

6.6 NUMERICAL METHOD

In what follows we assume that the data functions f_m and u_m are discrete functions in the interval $I = [0, 1]$, measured at the $N + 1$ sample points

$$x_i = ih, \quad i = 0, 1, \ldots, N, h = \frac{1}{N}.$$

Once the radius of mollification δ and the discrete function $J_\delta u_m$ are determined, we use centered differences to approximate $d/dx\, J_\delta u_m$ at the sample points of the interval $K_\delta = [3\delta, 1 - 3\delta]$. The number of sample points in K_δ is $M = \text{int}\{(1 - 6\delta)/h\} + 1$, and we consider the following partition of the set I:

$$x_0 = 0, \quad x_1 = 3\delta + h, \quad x_2 = 3\delta + 2h, \ldots, x_{M-2} = 1 - 3\delta - h, \quad x_{M-1} = 1.$$

Under assumption (A1), introducing the quantities

$$b_{\delta,\varepsilon}(x) = \alpha_{\delta,\varepsilon}(x)\frac{d}{dx}J_\delta u_m(x)$$

in (6.14), at the sample points of the interval I, we obtain

$$b_{\delta,\varepsilon}(x_i) = b_{\delta,\varepsilon}(x_{i-1}) + \int_{x_{i-1}}^{x_i} f_m(s)\, ds. \tag{6.22}$$

We define the finite-difference equation

$$B_{\delta,i}^\varepsilon = B_{\delta,i-1}^\varepsilon + hf_{m,i}, \quad i = 2, \ldots, M - 2, \tag{6.23a}$$

$$B_{\delta,1}^\varepsilon = c + (3\delta + h)f_{m,1}, \tag{6.23b}$$

$$B_{\delta,0}^\varepsilon = c, \tag{6.23c}$$

and begin by establishing a basic stability result for these coefficients.

Lemma 6.5 If f_m is uniformly Lipschitz on $I = [0, 1]$, with Lipschitz constant L, then

$$\|B_\delta^\varepsilon\|_{\infty, K_\delta} \le 2\|f_m\|_{\infty, I} + |c|$$

and

$$\|b_{\delta,\varepsilon} - B_\delta^\varepsilon\|_{\infty, K_\delta} \le L(9\delta^2 + 6\delta h + h).$$

Proof. According to (6.23), taking absolute values, we have

$$|B_{\delta,i}^\varepsilon| = |B_{\delta,i-1}^\varepsilon| + h|f_{m,i}| \le |B_{\delta,1}^\varepsilon| + (i-1)h\|f_m\|_{\infty, I}$$

$$\le |c| + (3\delta + h)\|f_m\|_{\infty, I} + (i-1)h\|f_m\|_{\infty, I}$$

$$\le |c| + 3\delta\|f_m\|_{\infty, I} + ih\|f_m\|_{\infty, I}$$

$$\le |c| + 2\|f_m\|_{\infty, I}, \quad i = 2, 3, \ldots, M - 2.$$

This proves the first part of the lemma.

Subtracting (6.23a) from (6.22), we obtain, for $2 \le i \le M - 2$,

$$
\begin{aligned}
\left| b_{\delta,\varepsilon}(x_i) - B^\varepsilon_{\delta,i} \right|
&\le \left| b_{\delta,\varepsilon}(x_{i-1}) - B^\varepsilon_{\delta,i-1} \right| + \int_{x_{i-1}}^{x_i} \left| f_m(s) - f_{m,i} \right| ds \\
&\le \left| b_{\delta,\varepsilon}(x_{i-1}) - B^\varepsilon_{\delta,i-1} \right| + Lh^2 \\
&\le (i - 1)h^2 L + \left| b_{\delta,\varepsilon}(x_1) - B^\varepsilon_{\delta,1} \right| \\
&\le (i - 1)h^2 L + L(x_1)^2 \\
&\le L(9\delta^2 + 6\delta h + h). \qquad \blacksquare
\end{aligned}
$$

In order to compute $\alpha_{\delta,\varepsilon}(x_i)$, we approximate $dJ_\delta u_m / dx(x_i)$ by centered differences. We introduce the notation

$$
P^\varepsilon_\delta(i) = \frac{J_\delta u_m(x_{i+1}) - J_\delta u_m(x_{i-1})}{2h}, \qquad 2 \le i \le M - 3,
$$

$$
P^\varepsilon_\delta(1) = \frac{J_\delta u_m(x_2) - J_\delta u_m(3\delta)}{2h},
$$

and

$$
P^\varepsilon_\delta(M - 2) = \frac{J_\delta u_m(1 - 3\delta) - J_\delta u_m(x_{M-3})}{2h}.
$$

These discrete approximations are such that

$$
\left| \frac{d}{dx} J_\delta u_m(x_i) - P^\varepsilon_\delta(i) \right| = O(h^2), \qquad i = 1, 2, \ldots, M - 2.
$$

We recall that under the conditions of Lemma 6.4,

$$
\left| \frac{d}{dx} J_\delta u_m(x) \right| \ge \lambda - \frac{2\varepsilon}{\delta \sqrt{\pi}}, \qquad x \in K_\delta.
$$

Thus, by the mean value theorem, we also have

$$
\left| P^\varepsilon_\delta(i) \right| \ge \lambda - \frac{2\varepsilon}{\delta \sqrt{\pi}}, \qquad i = 1, 2, \ldots, M - 2. \tag{6.24}
$$

The finite-difference approximations for the coefficients $\alpha_{\delta,\varepsilon}(x_i)$ are defined by

$$
A^\varepsilon_{\delta,i} = \frac{B^\varepsilon_{\delta,i}}{P^\varepsilon_\delta(i)}, \qquad i = 1, 2, \ldots, M - 2. \tag{6.25}
$$

The stability and convergence results for these coefficients are given in the following lemma.

Lemma 6.6 The finite-difference coefficients $A^\varepsilon_{\delta,i}$, $i = 1, 2, \ldots, M - 2$, satisfy

$$\|A^\varepsilon_\delta\|_{\infty, K_\delta} \leq \frac{\delta\sqrt{\pi}\left(|c| + 2\|f_m\|_{\infty, I}\right)}{\delta\lambda\sqrt{\pi} - 2\varepsilon}$$

and

$$\|\alpha_{\delta,\varepsilon} - A^\varepsilon_\delta\|_{\infty, K_\delta} \leq \frac{\delta\sqrt{\pi} L(9\delta^2 + 6\delta h + h)}{\lambda\delta\sqrt{\pi} - 2\varepsilon} + \frac{\left(|c| + 2\|f_m\|_{\infty, I}\right)\delta^2\pi\, O(h^2)}{\left(\delta\lambda\sqrt{\pi} - 2\varepsilon\right)^2}.$$

Proof. The first part of the lemma follows immediately from definition (6.25), inequality (6.24), and Lemma 6.5.

For the second part of the proof, we write

$$|\alpha_{\delta,\varepsilon}(x_i) - A^\varepsilon_{\delta,i}| = \left| \frac{b^\varepsilon_\delta(x_i)}{\dfrac{d}{dx}J_\delta u_m(x_i)} - \frac{B^\varepsilon_{\delta,i}}{P^\varepsilon_\delta(i)} \right|$$

$$\leq \frac{|b^\varepsilon_\delta(x_i) - B^\varepsilon_{\delta,i}|}{\left|\dfrac{d}{dx}J_\delta u_m(x_i)\right|} + |B^\varepsilon_{\delta,i}|\left|\frac{1}{\dfrac{d}{dx}J_\delta u_m(x_i)} - \frac{1}{P^\varepsilon_\delta(i)}\right|$$

$$\leq \frac{\delta\sqrt{\pi}\|b_{\delta,\varepsilon} - B^\varepsilon_\delta\|_{\infty, K_\delta}}{\lambda\delta\sqrt{\pi} - 2\varepsilon} + \frac{\left(|c| + 2\|f_m\|_{\infty, I}\right)\delta^2\pi\, O(h^2)}{\left(\delta\lambda\sqrt{\pi} - 2\varepsilon\right)^2}$$

$$\leq \frac{\delta\sqrt{\pi} L(9\delta^2 + 6\delta h + h)}{\lambda\delta\sqrt{\pi} - 2\varepsilon} + \frac{\left(|c| + 2\|f_m\|_{\infty, I}\right)\delta^2\pi\, O(h^2)}{\left(\delta\lambda\sqrt{\pi} - 2\varepsilon\right)^2},$$

using inequality (6.24), equation (6.25), and Lemma 6.5. ∎

Theorem 6.4 (Error Estimate) Under the conditions of Lemmas 6.3, 6.4, and 6.6,

$$\|\alpha - A^\varepsilon_\delta\|_{\infty, K_\delta} \leq C\delta\left[1 + \frac{\varepsilon(2 + \delta)}{\lambda\delta(\lambda\delta - 2\varepsilon)}\right]$$

$$+ \frac{\delta\sqrt{\pi} L(9\delta^2 + 6\delta h + h)}{\lambda\delta\sqrt{\pi} - 2\varepsilon} + \frac{\left(|c| + 2\|f_m\|_{\infty, I}\right)\delta^2\pi\, O(h^2)}{\left(\delta\lambda\sqrt{\pi} - 2\varepsilon\right)^2}.$$

Proof. It follows directly from Lemmas 6.3, 6.4, and 6.6 and the triangle inequality. ■

We now consider the case when u has a critical point at x_k, for some k. *Under assumption* (A2), the coefficients $\alpha_{\delta,\varepsilon}(x_i)$, $i = 1, 2, \ldots, M - 2$, $i \neq k$, are obtained from (6.20) and the corresponding $b_{\delta,\varepsilon}$ coefficients are given by

$$b_{\delta,\varepsilon}(x_i) = \int_{x_k}^{x_i} f(s)\, ds, \qquad i = 1, 2, \ldots, M - 2, i \neq k.$$

In this case, the finite-difference scheme is defined as follows:

$$B_{\delta,i}^\varepsilon = B_{\delta,i-1}^\varepsilon + h f_{m,i}, \qquad i = k + 1, k + 2, \ldots, M - 2, \quad (6.26a)$$

$$B_{\delta,i-1}^\varepsilon = B_{\delta,i}^\varepsilon + h f_{m,i}, \qquad i = k, k - 1, \ldots, 1, \quad\quad\quad (6.26b)$$

and

$$B_{\delta,k}^\varepsilon = 0. \tag{6.26c}$$

The basic stability results are obtained as before.

Lemma 6.7 If f_m is uniformly Lipschitz on $I = [0, 1]$, with Lipschitz constant L, then

$$\| B_\delta^\varepsilon \|_{\infty, K_\delta} \leq 2 \| f_m \|_{\infty, I}$$

and

$$\| b_{\delta,\varepsilon} - B_\delta^\varepsilon \|_{\infty, K_\delta} \leq L(9\delta^2 + 6\delta h + h).$$

Proof. Similar to the proof of Lemma 6.5, setting $c = 0$. ■

To approximate the coefficients $\alpha_{\delta,\varepsilon}(x_i)$, we consider the interval $V(x_k) = (x_k - h, x_k + h)$ and introduce

$$\lambda_h = \min\left\{ \left| \frac{du}{dx}(x_k - h) \right|, \left| \frac{du}{dx}(x_k + h) \right| \right\}.$$

We define

$$A_{\delta,i}^\varepsilon = \frac{B_{\delta,i}^\varepsilon}{P_\delta^\varepsilon(i)}, \qquad i = 2, 3, \ldots, M - 2, i \neq k, \tag{6.27a}$$

$$A_{\delta,k}^\varepsilon = \frac{A_{\delta,k+1}^\varepsilon + A_{\delta,k-1}^\varepsilon}{2}, \tag{6.27b}$$

with $P_\delta^\varepsilon(i)$ satisfying

$$P_\delta^\varepsilon(i) \geq \lambda_h - \frac{2\varepsilon}{\delta\sqrt{\pi}}, \qquad i \neq k. \tag{6.28}$$

For these finite-difference approximations, the stability result is as follows.

Lemma 6.8 The finite-difference coefficients $A_{\delta,i}^\varepsilon$, $i = 1, 2, \ldots, M - 2$, satisfy

$$\|A_\delta^\varepsilon\|_{\infty, K_\delta} \leq \frac{2\delta\sqrt{\pi}\,\|f_m\|_{\infty, I}}{\delta\lambda_h\sqrt{\pi} - 2\varepsilon}$$

and

$$\|\alpha_{\delta,\varepsilon} - A_\delta^\varepsilon\|_{\infty, K_\delta} \leq \frac{\delta\sqrt{\pi}\,L(9\delta^2 + 6\delta h + h)}{\lambda_h\delta\sqrt{\pi} - 2\varepsilon}$$
$$+ \frac{2\|f_m\|_{\infty, I}\,\delta^2\pi\,O(h^2)}{\left(\delta\lambda_h\sqrt{\pi} - 2\varepsilon\right)^2} + O(h^2).$$

Proof. Similar to Lemma 6.6. ■

Theorem 6.5 (Error Estimate)   Under the conditions of Lemmas 6.3, 6.4, and 6.8,

$$\|\alpha - A_\delta^\varepsilon\|_{\infty, K_\delta} \leq C\delta\left[1 + \frac{\varepsilon(2 + \delta)}{\lambda_h\delta(\lambda_h\delta - 2\varepsilon)}\right] + \frac{\delta\sqrt{\pi}\,L(9\delta^2 + 6\delta h + h)}{\lambda_h\delta\sqrt{\pi} - 2\varepsilon}$$
$$+ \frac{2\|f_m\|_{\infty, I}\,\delta^2\pi\,O(h^2)}{\left(\delta\lambda_h\sqrt{\pi} - 2\varepsilon\right)^2} + O(h^2).$$

Proof. Similar to Theorem 6.4. ■

6.7   NUMERICAL RESULTS

In this section we discuss the implementation of the method to some test problems of interest. In all the examples, the noisy data functions are obtained by adding an ε random error to f and u at the sample points of the interval $I = [0, 1]$, that is,

$$f_m(x_i) = f(x_i) + \varepsilon\theta_i,$$
$$u_m(x_i) = u(x_i) + \varepsilon\eta_i, \qquad i = 0, 1, \ldots, N,$$

where θ_i and η_i are Gaussian random variables with values in $[-1, 1]$. The selection of the radius of mollification is implemented as usual and the discrete weighted l^2 norm on K_δ is used to measure the solution error:

$$\| A_\delta^\varepsilon - \alpha \|_{2, K_\delta} = \left| \frac{1}{M-2} \sum_{j=1}^{M-2} \left[A_{\delta, j}^\varepsilon - \alpha(x_j) \right]^2 \right|^{1/2}.$$

EXAMPLE 3   In this example the exact data functions are

$$u(x) = x + \tfrac{1}{3}x^3,$$
$$f(x) = 4x(1 + x^2)$$

and the initial condition is given by

$$\alpha(0) \frac{du}{dx}(0) = 1.$$

The exact transmissivity coefficient is $\alpha(x) = 1 + x^2$. Table 6.1 shows the error norm of the solution as a function of δ and ε for $h = 0.01$. The qualitative behavior of a typical reconstructed transmissivity coefficient is shown in Fig. 6.3.

EXAMPLE 4   This is a nonlinear example. The exact data functions and the initial condition are

$$u(x) = e^x,$$
$$f(x) = 3e^{3x},$$
$$\alpha(u(0)) \frac{d}{dx}u(0) = 1.$$

The exact transmissivity coefficient is $\alpha(u) = u^2$. In this case, after computing $A_{\delta, i}^\varepsilon$, $i = 1, 2, \ldots, M - 2$, we collect all the pairs $(u_m(x_i), A_{\delta, i}^\varepsilon)$ to obtain the approximate functional relationship between the data function u and the transmissivity coefficient α. Table 6.2 illustrates the numerical stability of the method. A plot is shown in Fig. 6.4.

TABLE 6.1. Error Norm as a Function of ε in Example 3

ε	δ	Error Norm
0.000	0.04	0.0178
0.002	0.04	0.0195
0.005	0.04	0.0269

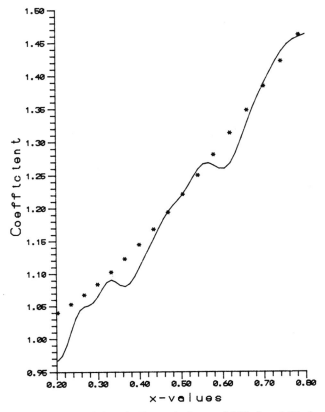

Fig. 6.3 Reconstructed coefficient in Example 3: $\varepsilon = 0.005$, $\delta = 0.04$, $\Delta x = h = 0.01$, $\|A - \alpha\| = 0.0269$. Exact ($***$); computed (—).

TABLE 6.2. Error Norm as a Function of ε in Example 4

ε	δ	Error Norm
0.000	0.04	0.0363
0.002	0.04	0.0372
0.005	0.06	0.0458

EXAMPLE 5   Here we consider the case where $d/dx\, u(x) = 0$ at a single point in the interval $I = [0, 1]$. The input data functions are

$$u(x) = -\sin \pi x,$$
$$f(x) = -\pi \cos \pi x + (1 + x)\pi^2 \sin \pi x.$$

The unique exact transmissivity coefficient to be reconstructed is $\alpha(x) =$

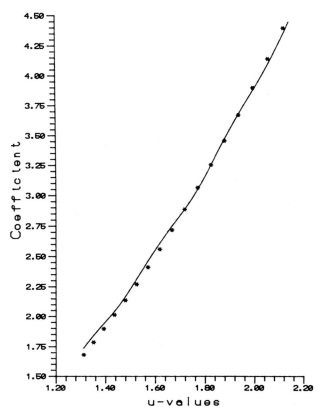

Fig. 6.4   Reconstructed coefficient in Example 4: $\varepsilon = 0.005$, $\delta = 0.06$, $\Delta x = h = 0.01$, $\|A - \alpha\| = 0.0329$. Exact ($\ast\ast\ast$); computed (—).

$1 + x$ and the singularity occurs at $x = \frac{1}{2}$, where $du/dx(\frac{1}{2}) = 0$. Figure 6.5 shows a graph of the reconstructed coefficient for $\varepsilon = 0.005$ and $\delta = 0.06$. The numerical stability properties of this example are summarized in Table 6.3.

Remark. We observe that in this case the initial condition is not needed; it is replaced by the "information" given at the singularity. In fact, once the centered differences $P_\delta^\varepsilon(i)$, $i = 1, 2, \ldots, M - 2$, have been computed and the singularity located at the node x_k, if $P_\delta^\varepsilon(k) \neq 0$, the discrepancy is used to correct (update) the values of $P_\delta^\varepsilon(i)$ in a neighborhood of the singularity. These updated values are then utilized in the evaluation of the approximated transmissivity coefficients $A_{\delta,i}^\varepsilon$ in (6.27) after determining $B_{\delta,i}^\varepsilon$, $i = 1, 2, \ldots, M - 2$, from (6.26).

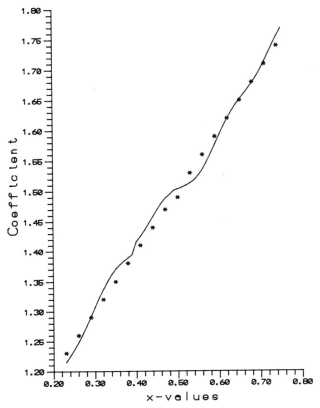

Fig. 6.5   Reconstructed coefficient in Example 5: $\varepsilon = 0.005$, $\delta = 0.06$, $\Delta x = h = 0.01$, $\|A - \alpha\| = 0.0361$. Exact ($* * *$); computed (—).

TABLE 6.3.   Error Norm as a Function of ε in Example 5

ε	δ	Error Norm
0.000	0.04	0.0212
0.002	0.06	0.0494
0.005	0.06	0.0502

6.8   IDENTIFICATION OF THE TRANSMISSIVITY COEFFICIENT IN THE ONE-DIMENSIONAL PARABOLIC EQUATION

The parabolic problem is essentially solved as a sequence of elliptic problems. For each time t, we apply the marching scheme in the x direction—as explained in the previous sections—and then repeat the process for a new updated time value.

The identification problem is now described as follows.

If the solution function $u(x, t)$ and the source function $f(x, t)$ are measured in the domain $I = [0, 1] \times [0, 1]$, identify the transmissivity coefficient $\alpha(x, t)$ such that

$$\begin{cases} \dfrac{\partial u}{\partial t}(x, t) + \dfrac{\partial}{\partial x}\left(\alpha(x, t)\dfrac{\partial u}{\partial x}(x, t)\right) = f(x, t), \\ \alpha(0, t)\dfrac{\partial u}{\partial x}(0, t) = g(t), \end{cases} \tag{6.29}$$

in the linear case, and

$$\begin{cases} \dfrac{\partial u}{\partial t}(x, t) + \dfrac{\partial}{\partial x}\left(\alpha(u(x, t))\dfrac{\partial u}{\partial x}(x, t)\right) = f(x, t), \\ \alpha(u(0, t))\dfrac{\partial u}{\partial x}(0, t) = g(t), \end{cases} \tag{6.30}$$

in the nonlinear case.

As in the elliptic case, we study these two problems under each of the following assumptions:

$$\inf_{K}\left\{\left\|\dfrac{\partial u}{\partial x}(x, t)\right\|\right\} > 0 \tag{A1'}$$

or

$$\inf_{x \in K_t}\left\{\max\left(\left|\dfrac{\partial u}{\partial x}(x, t)\right|, \dfrac{\partial^2 u}{\partial x^2}(x, t)\right)\right\} > 0, \tag{A2'}$$

where the arguments (x, t) are taken in some suitable compact set $K \subset I$ and, for each t, $K_t = \{x: (x, t) \in K\}$. We also assume

$$\|f - f_m\|_{\infty, I} \le \varepsilon, \qquad \|u - u_m\|_{\infty, I} \le \varepsilon, \qquad \text{and} \qquad \|g - g_m\|_{\infty, I} \le \varepsilon. \tag{6.31}$$

6.9 STABILITY ANALYSIS

Under assumption $(A1')$, (6.29) and (6.30) can be integrated with respect to x, obtaining the unique solutions

$$\alpha(x, t) = \dfrac{\displaystyle\int_0^x\left(f(s, t) - \dfrac{\partial u}{\partial t}(s, t)\right) ds + g(t)}{\dfrac{\partial u}{\partial x}(x, t)} \tag{6.32}$$

and

$$\alpha(u(x,t)) = \frac{\int_0^x \left(f(s,t) - \frac{\partial u}{\partial t}(s,t) \right) ds + g(t)}{\frac{\partial u}{\partial x}(x,t)}, \quad (6.33)$$

respectively. Once again, both formulas are similar, and, consequently, we devote our attention to the linear case.

The approximate transmissivity coefficients $\alpha_{\delta_1\delta_2}(x,t)$ and $\alpha_{\delta_1\delta_2,\varepsilon}(x,t)$ are respectively defined by

$$\alpha_{\delta_1\delta_2}(x,t) = \frac{\int_0^x \left(f(s,t) - \frac{\partial J_{\delta_2}u}{\partial t}(s,t) \right) ds + g(t)}{\frac{\partial J_{\delta_1}u}{\partial x}(x,t)} \quad (6.34)$$

and

$$\alpha_{\delta_1\delta_2,\varepsilon}(x,t) = \frac{\int_0^x \left(f_m(s,t) - \frac{\partial J_{\delta_2}u_m}{\partial t}(s,t) \right) ds + g_m(t)}{\frac{\partial J_{\delta_1}u_m}{\partial x}(x,t)}. \quad (6.35)$$

Here, $\partial J_{\delta_2}u/\partial t$ and $\partial J_{\delta_1}u/\partial x$ indicate one-dimensional mollifications in the t and x directions, respectively, and, in order to simplify some calculations, in what follows we write α_δ and $\alpha_{\delta,\varepsilon}$ with $\delta = \max(\delta_1, \delta_2)$.

For the parabolic identification problem, the preliminary results are as follows.

Lemma 6.9 (Consistency)   If

$$\max\left(\left| \frac{\partial^2 u}{\partial x^2}(x,t) \right|, \left| \frac{\partial^2 u}{\partial t^2}(x,t) \right| \right) \le M_2$$

and

$$\left| \frac{\partial u}{\partial x}(x,t) \right| \ge \lambda > 0, \ (x,t) \in K_\delta \subset I,$$

then

$$\|\alpha - \alpha_\delta\|_{\infty, K_\delta} \le C\delta, \quad (6.36)$$

for some constant $C = C(\lambda, M_2, \|\nabla u\|_{\infty, K_\delta}, \|f\|_{\infty, I}, \|g\|_{\infty, I})$.

Proof. From (6.32) and (6.34), for $(x, t) \in K_\delta$, we have

$$|\alpha(x, t) - \alpha_\delta(x, t)| \leq \frac{1}{\lambda^2} \left| \left[\int_0^x \left(f(s, t) - \frac{\partial u}{\partial t}(s, t) \right) ds + g(t) \right] \frac{\partial J_\delta u}{\partial x}(x, t) \right.$$

$$\left. - \left[\int_0^x \left(f(s, t) - \frac{\partial J_\delta u}{\partial t}(s, t) \right) ds + g(t) \right] \frac{\partial u}{\partial x}(x, t) \right|$$

$$\leq \frac{1}{\lambda^2} \left[\int_0^x |f(s, t)| \, ds \left| \frac{\partial J_\delta u}{\partial x}(x, t) - \frac{\partial u}{\partial x}(x, t) \right| \right.$$

$$+ \int_0^x \left| \frac{\partial J_\delta u}{\partial t}(s, t) - \frac{\partial u}{\partial t}(s, t) \right| ds \left| \frac{\partial u}{\partial x}(x, t) \right|$$

$$+ \int_0^x \left| \frac{\partial u}{\partial t}(s, t) \right| ds \left| \frac{\partial J_\delta u}{\partial x}(x, t) - \frac{\partial u}{\partial x}(x, t) \right|$$

$$\left. + |g(t)| \left| \frac{\partial J_\delta u}{\partial x}(x, t) - \frac{\partial u}{\partial x}(x, t) \right| \right]$$

$$\leq \frac{1}{\lambda^2} \left[\left\| \frac{\partial J_\delta u}{\partial x} - \frac{\partial u}{\partial x} \right\|_{\infty, K_\delta} \left(\|f\|_{\infty, K_\delta} + \left\| \frac{\partial u}{\partial t} \right\|_{\infty, K_\delta} + \|g\|_{\infty, [0, 1]} \right) \right.$$

$$\left. + \left\| \frac{\partial J_\delta u}{\partial t} - \frac{\partial u}{\partial t} \right\|_{\infty, K_\delta} \left\| \frac{\partial u}{\partial x} \right\|_{\infty, K_\delta} \right]$$

$$\leq \frac{3 M_2 \delta}{\lambda^2} \left(\|f\|_{\infty, K_\delta} + 2\|\nabla u\|_{\infty, K_\delta} + \|g\|_{\infty, [0, 1]} \right)$$

$$\leq C\delta$$

by Lemma 1.1, with $C = C(\lambda, M_2, \|f\|_{\infty, K_\delta}, \|\nabla u\|_{\infty, K_\delta}, \|g\|_{\infty, [0, 1]})$. ∎

Lemma 6.10 (Stability) If u_m, g_m, and f_m verify inequalities (6.31),

$$\left| \frac{\partial u}{\partial x}(x, t) \right| \geq \lambda > 0, \qquad (x, t) \in K_\delta \subset I, \qquad \text{and} \qquad \delta > \frac{2\varepsilon}{\lambda\sqrt{\pi}},$$

then

$$\|\alpha_\delta - \alpha_{\delta, \varepsilon}\|_{\infty, K_\delta} \leq \frac{C\varepsilon}{\lambda(\lambda\delta\sqrt{\pi} - 2\varepsilon)} (\lambda\delta + 1), \tag{6.37}$$

where $C = C(M_2, \|f\|_{\infty, K_\delta}, \|\nabla u\|_{\infty, K_\delta}, \|g\|_{\infty, [0, 1]})$.

Proof. We leave this as an exercise. ∎

Combining the last two lemmas, we obtain the following result.

Theorem 6.6 (Error Estimate)   Under the conditions of Lemmas 6.9 and 6.10,

$$\|\alpha - \alpha_{\delta,\varepsilon}\|_{\infty, K_\delta} \le C\left(\delta + \frac{\varepsilon(\lambda\delta + 1)}{\lambda(\lambda\delta\sqrt{\pi} - 2\varepsilon)}\right). \tag{6.38}$$

Remark. Similar bounds for the approximate transmissivity coefficient $\alpha_{\delta,\varepsilon}$ are obtained in the nonlinear case.

Under assumption (A2′), where for each fixed time t, $t_{\min} \le t \le t_{\max}$, u has at most one critical point (x^*, t) in K_δ, for $(x, t) \in K_\delta$, $(x, t) \ne (x^*, t)$, we have

$$\alpha(x, t) = \frac{\int_{x^*}^x \left[f(s, t) - \frac{\partial u}{\partial t}(s, t)\right] ds + g(t)}{\frac{\partial u}{\partial x}(x, t)}, \tag{6.39}$$

$$\alpha_\delta(x, t) = \frac{\int_{x^*}^x \left[f(s, t) - \frac{\partial J_\delta u}{\partial t}(s, t)\right] ds + g(t)}{\frac{\partial J_\delta u}{\partial x}(x, t)}, \tag{6.40}$$

and

$$\alpha_{\delta,\varepsilon}(x, t) = \frac{\int_0^x \left[f_m(s, t) - \frac{\partial J_\delta u_m}{\partial t}(s, t)\right] ds + g_m(t)}{\frac{\partial J_\delta u_m}{\partial x}(x, t)} \tag{6.41}$$

in the linear case.
    Also in the linear case, at $(x, t) = (x^*, t)$, we have

$$\alpha(x^*, t) = \frac{f(x^*, t) - \frac{\partial u}{\partial t}(x^*, t)}{\frac{\partial^2 u}{\partial x^2}(x^*, t)}. \tag{6.42}$$

For the nonlinear case we proceed similarly.
    We observe that under assumption (A2′) no boundary conditions are needed. Moreover, for each fixed t, we can select an open interval $V_t(x^*) =$

$\{(x, t): x^* - \gamma < x < x^* + \gamma\}$, such that $|\partial u/\partial x(x, t)| \geq \lambda_{x^*} > 0$ for all (x, t) $\in V_t(x^*) \cap K_\delta$.

We will explain later how to approximate the value of α at any point in $V_t(x^*)$ and avoid the computation of the second derivative in (6.42).

6.10   NUMERICAL METHOD

In what follows we assume that the data functions f_m and u_m are discrete functions in the interval $I = [0, 1] \times [0, 1]$, measured at the $(N_1 + 1)(N_2 + 1)$ sample points

$$(x_i, t_k) = (ih_1, kh_2),$$

$$i = 0, 1, \ldots, N_1, \ k = 0, 1, \ldots, N_2, \ h_1 = \frac{1}{N_1}, \ h_2 = \frac{1}{N_2}.$$

Once the radii of mollification δ_1, δ_2 and the discrete function $J_\delta u_m$ for $\delta = \max(\delta_1, \delta_2)$ are determined in the x and t directions, we use centered differences to approximate the partial derivatives of $J_\delta u_m$ at the sample points of the compact set $K_\delta = [3\delta, 1 - 3\delta] \times [3\delta, 1 - 3\delta]$. In the sequel we set $h_1 = h_2 = h$ to simplify some analysis, and we consider the partition of the set I defined by the pairs (x_i, t_k), $0 \leq i, k \leq M - 1$ such that

$$x_0 = 0, \quad x_1 = 3\delta + h, \quad x_2 = 3\delta + 2h, \ldots, x_{M-2} = 1 - 3\delta - h,$$
$$x_{M-1} = 1, \quad t_0 = 0, \quad t_1 = 3\delta + h,$$
$$t_2 = 3\delta + 2h, \ldots, t_{M-2} = 1 - 3\delta - h, \quad t_{M-1} = 1.$$

Under assumption (A1'), introducing the quantities

$$b_{\delta, \varepsilon}(x, t) = \alpha_{\delta, \varepsilon}(x, t) \frac{\partial J_\delta u_m}{\partial x}(x, t)$$

in (6.35), at the sample points of the set I, we obtain

$$b_{\delta, \varepsilon}(x_i, t_k) = b_{\delta, \varepsilon}(x_{i-1}, t_k) + \int_{x_{i-1}}^{x_i} \left[f_m(s, t_k) - \frac{\partial J_\delta u_m}{\partial t}(s, t_k) \right] ds. \quad (6.43)$$

We define the finite-difference equation

$$B_{\delta, i, k}^\varepsilon = B_{\delta, i-1, k}^\varepsilon + h(f_{m, i, k} - T_\delta^\varepsilon(i, k)), \quad i = 2, \ldots, M - 2, \quad (6.44a)$$
$$B_{\delta, 1, k}^\varepsilon = g_{m, k} + (3\delta + h)(f_{m, 1} - T_\delta^\varepsilon(1, k)), \quad (6.44b)$$
$$B_{\delta, 0}^\varepsilon = g_{m, k}, \quad (6.44c)$$

where $T_\delta^\varepsilon(i, k)$ is the central-difference approximation to $\partial J_\delta u_m / \partial t (x_i, t_k)$ and satisfies

$$\left| T_\delta^\varepsilon(i, k) - \frac{\partial J_\delta u_m}{\partial t}(x_i, t_k) \right| = O(h^2).$$

We now establish a basic stability result for these coefficients.

Lemma 6.11   If f_m and u_m are uniformly Lipschitz in the x direction on $I_x = [0, 1]$, with Lipschitz constants L_1 and L_2, respectively, then

$$\| B_\delta^\varepsilon \|_{\infty, K_\delta} \le 2\left(\| f_m \|_{\infty, I} + \left\| \frac{\partial J_\delta u_m}{\partial t} \right\|_{\infty, K_\delta} \right) + \| g \|_{\infty, [0, 1]}$$

and

$$\| b_{\delta, \varepsilon} - B_\delta^\varepsilon \|_{\infty, K_\delta} \le \left(L_1 + 8\frac{L_2}{\delta} \right)(9\delta^2 + 6\delta h + h) + O(h^2).$$

Proof.   According to (6.44), taking absolute values, we have, for each $i = 1, 2, \ldots, M - 2$ and $k = 1, 2, \ldots, M - 2$,

$$\begin{aligned}
|B_{\delta, i, k}^\varepsilon| &= |B_{\delta, i-1, k}^\varepsilon + h| f_{m, i, k} - T_\delta^\varepsilon(i, k)| \\
&\le |B_{\delta, 1, k}^\varepsilon| + (i - 1)h\left(\| f_m \|_{\infty, I} + \| T_\delta^\varepsilon \|_{\infty, K_\delta} \right) \\
&\le \| g_m \|_{\infty, [0, 1]} + (3\delta + h)\left(\| f_m \|_{\infty, I} + \| T_\delta^\varepsilon \|_{\infty, K_\delta} \right) \\
&\quad + (i - 1)h\left(\| f_m \|_{\infty, I} + \| T_\delta^\varepsilon \|_{\infty, K_\delta} \right) \\
&\le \| g_m \|_{\infty, [0, 1]} + 2\left(\| f_m \|_{\infty, I} + \left\| \frac{\partial J_\delta u_m}{\partial t} \right\|_{\infty, K_\delta} \right)
\end{aligned}$$

by the mean value theorem.

This proves the first part of the lemma.

Subtracting (6.44a) from (6.43), we obtain, for $2 \le i \le M - 2$,

$$\begin{aligned}
|b_{\delta, \varepsilon}(x_i, t_k) - B_{\delta, i, k}^\varepsilon| &\le |b_{\delta, \varepsilon}(x_{i-1}, t_k) - B_{\delta, i-1, k}^\varepsilon| \\
&\quad + \int_{x_{i-1}}^{x_i} |f_m(s, t_k) - f_m(x_i, t_k)| \, ds \\
&\quad + \int_{x_{i-1}}^{x_i} \left| \frac{\partial J_\delta u_m}{\partial t}(s, t_k) - \frac{\partial J_\delta u_m}{\partial t}(x_i, t_k) \right| ds + O(h^3) \\
&\le |b_{\delta, \varepsilon}(x_{i-1}, t_k) - B_{\delta, i-1, k}^\varepsilon| + L_1 h^2 \\
&\quad + \int_{x_{i-1}}^{x_i} \left| \frac{\partial J_\delta u_m}{\partial t}(s, t_k) - \frac{\partial J_\delta u_m}{\partial t}(x_i, t_k) \right| ds + O(h^3).
\end{aligned}$$

We estimate the integral term as follows.

$$\int_{x_{i-1}}^{x_i} \left| \frac{\partial J_\delta u_m}{\partial t}(s, t_k) - \frac{\partial J_\delta u_m}{\partial t}(x_i, t_k) \right| ds$$

$$\leq \int_{x_{i-1}}^{x_i} \left\{ \int_{-\infty}^{\infty} \left| \frac{\partial}{\partial t} \rho_\delta(t_k - z) \right| |u_m(s, z) - u_m(x_i, z)| \, dz \right\} ds$$

$$\leq \int_{x_{i-1}}^{x_i} \left\{ L_2 |s - x_i| \int_{-\infty}^{\infty} \left| \frac{\partial}{\partial t} \rho_\delta(t_k - z) \right| dz \right\} ds \leq L_2 h^2 \frac{8}{\delta}.$$

Therefore,

$$\left| b_{\delta, \varepsilon}(x_i, t_k) - B_{\delta, i, k}^\varepsilon \right|$$

$$\leq \left| b_{\delta, \varepsilon}(x_{i-1}, t_k) - B_{\delta, i-1, k}^\varepsilon \right| + L_1 h^2 + L_2 h^2 \frac{8}{\delta} + O(h^3)$$

$$\leq \left| b_{\delta, \varepsilon}(x_1, t_k) - B_{\delta, 1, k}^\varepsilon \right| + (i - 1)h^2 \left(L_1 + \frac{8}{\delta} L_2 \right) + O(h^2)$$

$$\leq \left(L_1 + \frac{8}{\delta} L_2 \right)(9\delta^2 + 6\delta h + h) + O(h^2). \quad \blacksquare$$

In order to compute $\alpha_{\delta, \varepsilon}(x_i, t_k)$, we approximate $\partial J_\delta u / \partial x(x_i, t_k)$ by centered differences. We introduce the notation

$$P_\delta^\varepsilon(i, k) = \frac{J_\delta u_m(x_{i+1}, t_k) - J_\delta u_m(x_{i-1}, t_k)}{2h}, \qquad 2 \leq i \leq M - 3,$$

$$P_\delta^\varepsilon(1, k) = \frac{J_\delta u_m(x_2, t_k) - J_\delta u_m(3\delta, t_k)}{2h},$$

and

$$P_\delta^\varepsilon(M - 2, k) = \frac{J_\delta u_m(1 - 3\delta, t_k) - J_\delta u_m(x_{M-3}, t_k)}{2h}, \qquad 1 \leq k \leq M - 2.$$

These discrete approximations are such that

$$\left| \frac{\partial J_\delta u_m}{\partial x}(x_i, t_k) - P_\delta^\varepsilon(i, k) \right| = O(h^2), \qquad 1 \leq i, k \leq M - 2.$$

We recall that under the conditions of Lemma 6.4, for fixed t,

$$\left| \frac{\partial J_\delta u_m}{\partial x}(x,t) \right| \geq \lambda - \frac{2\varepsilon}{\delta\sqrt{\pi}}, \qquad (x,t) \in K_\delta.$$

Thus, by the mean value theorem, we also have for each k, $1 \leq k \leq M - 2$,

$$|P_\delta^\varepsilon(i,k)| \geq \lambda - \frac{2\varepsilon}{\delta\sqrt{\pi}}, \qquad 1 \leq i \leq M - 2. \tag{6.45}$$

The finite-difference approximations for the coefficients $\alpha_{\delta,\varepsilon}(x_i, t_k)$ are defined by

$$A_{\delta,i,k}^\varepsilon = \frac{B_{\delta,i,k}^\varepsilon}{P_\delta^\varepsilon(i,k)}, \qquad 1 \leq i, k \leq M - 2. \tag{6.46}$$

The stability and convergence results for these coefficients are given in the following lemma.

Lemma 6.12 The finite-difference coefficients $A_{\delta,i,k}^\varepsilon$, $1 \leq i$, $k \leq M - 2$, satisfy

$$\|A_\delta^\varepsilon\|_{\infty, K_\delta} \leq \frac{\delta\sqrt{\pi}\left(\|g_m\|_{\infty,[0,1]} + 2\|f_m\|_{\infty,I} + \left\| \frac{\partial J_\delta u_m}{\partial x} \right\|_{\infty,K_\delta} \right)}{\delta\lambda\sqrt{\pi} - 2\varepsilon}$$

and

$$\|\alpha_{\delta,\varepsilon} - A_\delta^\varepsilon\|_{\infty, K_\delta} \leq \frac{\delta\sqrt{\pi}\left[\left(L_1 + \frac{8}{\delta}L_2 \right)(9\delta^2 + 6\delta h + h) + O(h^2) \right]}{\lambda\delta\sqrt{\pi} - 2\varepsilon}$$

$$+ \frac{\left[\|g\|_{\infty,[0,1]} + 2\left(\|f_m\|_{\infty,I} + \left\| \frac{\partial J_\delta u_m}{\partial x} \right\|_{\infty,K_\delta} \right) \right] \delta^2\pi\, O(h^2)}{\left(\delta\lambda\sqrt{\pi} - 2\varepsilon \right)^2}.$$

Proof. The first part of the lemma follows immediately from definition (6.46), inequality (6.45), and Lemma 6.11.

For the second part of the proof, we write

$$
\left| \alpha_{\delta,\varepsilon}(x_i, t_k) - A^\varepsilon_{\delta,i,k} \right|
$$

$$
= \left| \frac{b_{\delta,\varepsilon}(x_i, t_k)}{\dfrac{\partial J_\delta u_m}{\partial x}(x_i, t_k)} - \frac{B^\varepsilon_{\delta,i,k}}{P^\varepsilon_\delta(i,k)} \right|
$$

$$
= \frac{\left| b_{\delta,\varepsilon}(x_i, t_k) - B^\varepsilon_{\delta,i,k} \right|}{\left| \dfrac{\partial J_\delta u_m}{\partial x}(x_i, t_k) \right|} + \left| B^\varepsilon_{\delta,i,k} \right| \left| \frac{1}{\dfrac{\partial J_\delta u_m}{\partial x}(x_i, t_k)} - \frac{1}{P^\varepsilon_\delta(i,k)} \right|
$$

$$
\leq \frac{\delta\sqrt{\pi}\, \|b_{\delta,\varepsilon} - B^\varepsilon_\delta\|_{\infty, K_\delta}}{\lambda \delta \sqrt{\pi} - 2\varepsilon}
$$

$$
+ \frac{\left[\|g\|_{\infty,[0,1]} + 2\left(\|f_m\|_{\infty, I} + \left\| \dfrac{\partial J_\delta u_m}{\partial x} \right\|_{\infty, K_\delta} \right) \right] \delta^2 \left\| \dfrac{\partial J_\delta u_m}{\partial x} - P^\varepsilon_\delta \right\|_{\infty, K_\delta}}{\left(\delta \lambda \sqrt{\pi} - 2\varepsilon \right)^2}
$$

$$
\leq \frac{\delta\sqrt{\pi}\left[\left(L_1 + \dfrac{8}{\delta} L_2 \right)(9\delta^2 + 6\delta h + h) + O(h^2) \right]}{\lambda \delta \sqrt{\pi} - 2\varepsilon}
$$

$$
+ \frac{\left[\|g\|_{\infty,[0,1]} + 2\left(\|f_m\|_{\infty, I} + \left\| \dfrac{\partial J_\delta u_m}{\partial x} \right\|_{\infty, K_\delta} \right) \right] \delta^2 \pi\, O(h^2)}{\left(\delta \lambda \sqrt{\pi} - 2\varepsilon \right)^2},
$$

using (6.45) and (6.46) and Lemma 6.11. ■

Theorem 6.7 (Error Estimate) Under the conditions of Lemmas 6.9, 6.10, and 6.12,

$$
\|\alpha - A^\varepsilon_\delta\|_{\infty, K_\delta} \leq C\left(\delta + \frac{\varepsilon(\lambda\delta + 1)}{\lambda(\lambda\delta\sqrt{\pi} - 2\varepsilon)} \right)
$$

$$
+ \frac{\delta\sqrt{\pi}\left[\left(L_1 + \dfrac{8}{\delta} L_2 \right)(9\delta^2 + 6\delta h + h) + O(h^2) \right]}{\lambda \delta \sqrt{\pi} - 2\varepsilon}
$$

$$
+ \frac{\left[\|g\|_{\infty,[0,1]} + 2\left(\|f_m\|_{\infty, I} + \left\| \dfrac{\partial J_\delta u_m}{\partial x} \right\|_{\infty, K_\delta} \right) \right] \delta^2 \pi\, O(h^2)}{\left(\delta \lambda \sqrt{\pi} - 2\varepsilon \right)^2}.
$$

Proof. It follows directly from Lemmas 6.9, 6.10, and 6.12 and the triangle inequality. ■

Under assumption (A2'), if $\partial J_\delta u_m / \partial x(x_{i_0}, t_k) = 0$ for $(x_{i_0}, t_k) \in K_\delta$, the finite-difference scheme is defined as follows:

$$B_{\delta, i, k}^\varepsilon = B_{\delta, i-1, k}^\varepsilon + h(f_{m, i, k} - P_\delta^\varepsilon(i, k)),$$
$$i = i_0 + 1, i_0 + 2, \ldots, M - 2, \quad (6.47a)$$

$$B_{\delta, i-1, k}^\varepsilon = B_{\delta, i, k}^\varepsilon + h(f_{m, i, k} - P_\delta^\varepsilon(i, k)), \quad i = i_0, i_0 - 1, \ldots, 2, \quad (6.47b)$$

and

$$B_{\delta, i_0, k} = 0. \quad (6.47c)$$

The basic stability results are obtained as before.

To approximate the coefficients $\alpha_{\delta, \varepsilon}(x_i, t_k)$, we consider for each fixed time t_k, the interval $V_{t_k}(x_{i_0}) = \{(x, t_k): x_{i_0} - h \le x \le x_{i_0} + h\}$ and introduce

$$\lambda_{h, k} = \min\left\{ \left| \frac{\partial u}{\partial x}(x_{i_0} - h, t_k) \right|, \left| \frac{\partial u}{\partial x}(x_{i_0} + h, t_k) \right| \right\}.$$

We define

$$A_{\delta, i, k}^\varepsilon = \frac{B_{\delta, i, k}^\varepsilon}{P_\delta^\varepsilon(i, k)}, \quad i \ne i_0, \quad (6.48a)$$

$$A_{\delta, i_0, k}^\varepsilon = \frac{A_{\delta, i_0+1, k}^\varepsilon + A_{\delta, i_0-1, k}^\varepsilon}{2}, \quad (6.48b)$$

with $P_\delta^\varepsilon(i, k)$ satisfying

$$\left| P_\delta^\varepsilon(i, k) \right| \ge \lambda_{h, k} - \frac{2\varepsilon}{\delta \sqrt{\pi}}, \quad i \ne i_0. \quad (6.49)$$

For these finite-difference approximations, the stability and convergence results are similar to the ones stated in Lemmas 6.9 and 6.10 and Theorem 6.7.

6.11 NUMERICAL RESULTS

In this section we discuss the implementation of the method by numerically solving some test problems of interest over the unit square. In all the

TABLE 6.4. Error Norm as a Function of ε in Example 6

ε	δ	Error Norm
0.000	0.04	0.0001
0.002	0.06	0.0065
0.005	0.06	0.0162

examples, the noisy data functions are obtained by adding an ε random error to f and u at the sample points of the square $I = [0, 1] \times [0, 1]$, that is,

$$f_m(x_i, t_k) = f(x_i, t_k) + \varepsilon\theta_{i, k},$$
$$u_m(x_i, t_k) = u(x_i, t_k) + \varepsilon\eta_{i, k},$$
$$0 \leq i \leq N_1, 0 \leq k \leq N_2,$$

where $\theta_{i, k}$ and $\eta_{i, k}$ are Gaussian random variables with values in $[-1, 1]$. The selection of the radius of mollification is implemented as usual and with $N_1 = N_2 = N$, the discrete weighted l^2 norm on K_δ, $\delta = \max(\delta_1, \delta_2)$, is used to measure the solution error. Thus, we have

$$\| A_\delta^\varepsilon - \alpha \|_{2, K_\delta} = \left| \frac{1}{(M - 2)^2} \sum_{n=1}^{M-2} \sum_{j=1}^{M-2} \left[A_{\delta, j, n}^\varepsilon - \alpha(x_j, t_n) \right]^2 \right|^{1/2}.$$

EXAMPLE 6   In this example the exact data functions are

$$u(x, t) = xt,$$
$$f(x, t) = x$$

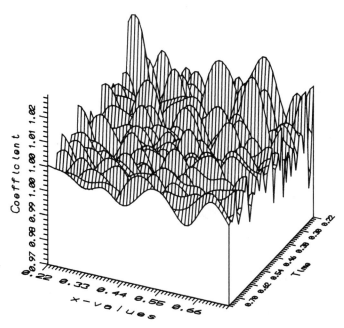

Fig. 6.6 Reconstructed transmissivity coefficient in Example 6: $\varepsilon = 0.005$, $\delta = 0.06$, $\Delta t = \Delta x = 0.01$.

and the initial condition is given by

$$\alpha(0,t)\frac{\partial u}{\partial x}(0,t) = t.$$

The exact transmissivity coefficient is $\alpha(x,t) = 1$ and the parabolic equation reads

$$\frac{\partial u}{\partial t} + \frac{\partial^2 u}{\partial x^2} = x, \qquad 0 < x < 1, 0 < t < 1,$$

with exact solution $u(x,t) = xt$.

Table 6.4 shows the error norm of the solution as a function of δ and ε for $h_1 = h_2 = h = 0.01$. Figure 6.6 illustrates the qualitative behavior of the numerical solution for $\varepsilon = 0.005$ and $\delta = 0.06$.

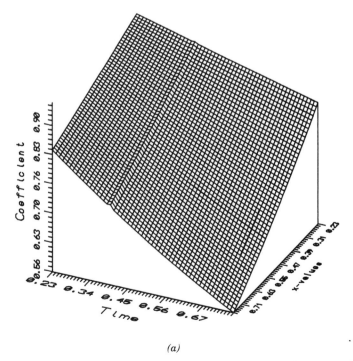

(a)

Fig. 6.7a Exact transmissivity coefficient in Example 7: singularity at $x = 0.4$.

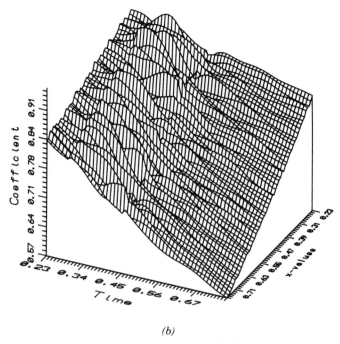

(b)

Fig. 6.7b   Reconstructed transmissivity coefficient in Example 7: $\varepsilon = 0.005$, $\delta = 0.06$, $\Delta t = \Delta x = 0.01$; singularity at $x = 0.4$.

EXAMPLE 7   Here we consider the case where $\partial u / \partial x(x_{i_0}, t) = 0$ for all times t. The differential equation is given by

$$\frac{\partial u}{\partial t} + \frac{\partial}{\partial x}\left(e^{-xt}\frac{\partial u}{\partial x}\right) = 2\left[-te^{-t}(x - 0.4) + (t - 0.4) + e^{-xt}\right],$$

$$0 < x < 1, 0 < t < 1,$$

with exact solution $u(x, t) = (x - 0.4)^2 + (t - 0.4)^2$.
    The exact input data functions are

$$u(x, t) = (x - 0.4)^2 + (t - 0.4)^2,$$

$$f(x, t) = 2\left[-te^{-t}(x - 0.4) + (t - 0.4) + e^{-xt}\right].$$

The unique exact transmissivity coefficient to be reconstructed is $\alpha(x, t) = e^{-xt}$ and the singularity occurs at $x_{i_0} = 0.4$. Figures 6.7a and b show graphs of the exact transmissivity coefficient and the reconstructed coefficient for $\varepsilon = 0.005$ and $\delta = 0.06$, respectively. The numerical stability properties of this example are summarized in Table 6.5 for $h_1 = h_2 = h = 0.01$. The

TABLE 6.5. Error Norm as a Function of ε in Example 7

ε	δ	Error Norm
0.000	0.04	0.0051
0.002	0.06	0.0228
0.005	0.06	0.0445

remark at the end of Sec. 6.7—singular elliptic case—applies also to this parabolic example: for each value of t, the starting information for the marching scheme is obtained at the location of the singularity and no initial condition at the boundary is needed.

EXAMPLE 8   This is a nonlinear example. The exact data functions and the initial condition are

$$u(x,t) = xt,$$

$$f(x,t) = x + t^2,$$

$$\alpha(u(0,t))\frac{\partial u}{\partial x}(0,t) = 0,$$

corresponding to the partial differential equation

$$\frac{\partial u}{\partial t} + \frac{\partial}{\partial x}\left(u\frac{\partial u}{\partial x}\right) = x + t^2, \qquad 0 < x < 1, 0 < t < 1,$$

with solution $u(x,t) = xt$.

The exact transmissivity coefficient is $\alpha(u) = u$. In this case, after computing $A^{\varepsilon}_{\delta,i,k}$, $1 \le i$, $k \le M - 2$, we collect all the pairs $(A^{\varepsilon}_{\delta,i,k}, u_m(x_i, t_k))$—about 4000 with $h = 0.01$ and $\delta = 0.06$—to obtain the approximate functional relationship between the data function u and the transmissivity coefficient α. A plot is shown in Fig. 6.8 corresponding to the noise level $\varepsilon = 0.005$. Table 6.6 illustrates the numerical stability of the method.

6.12   EXERCISES

6.1. Consider the initial value problem

$$\frac{d^2y}{dt^2}(t) = y(t) + p(t), \qquad 0 < t < 1,$$

$$y(0) = 1, \qquad \frac{dy}{dt}(0) = 1 - 5\pi, \qquad \text{and} \qquad p(t) = (1 + 25\pi^2)\sin 5\pi t,$$

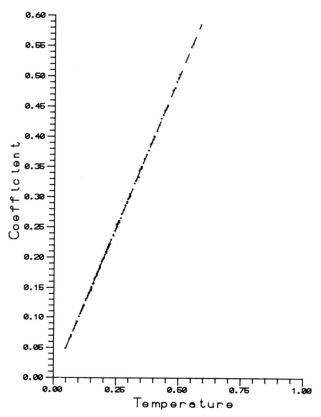

Fig. 6.8 Reconstructed transmissivity coefficient in Example 8: $\varepsilon = 0.005$, $\delta = 0.06$, $\Delta t = \Delta x = 0.01$.

TABLE 6.6. Error Norm as a Function of ε in Example 8

ε	δ	Error Norm
0.000	0.04	0.0030
0.002	0.06	0.0033
0.005	0.06	0.0048

which has exact solution $y(t) = e^t - \sin 5\pi t$, $0 \leq t \leq 1$. Using the mollification method, numerically identify the forcing term $p(t)$ by

(a) reducing the second-order differential equation to a first-order system.

(b) directly approximating the second derivative. Consider as data for the inverse problem the exact solution $y(t)$ perturbed with random noise of amplitude $\varepsilon = 0.005$ and sample step $\Delta t = 0.01$. Compare the relative error norms in both cases.

6.2. Prove Lemma 6.10.

6.3. Introduce

$$\rho_\delta(\mathbf{x}) = \rho_{\delta_1}(\mathbf{x}_1)\rho_{\delta_2}(\mathbf{x}_2) \cdots \rho_{\delta_n}(\mathbf{x}_n),$$

where

$$\rho_{\delta_i}(\mathbf{x}_i) = \frac{1}{\delta_i\sqrt{\pi}} \exp\left[-\left(\frac{x_i}{\delta_i}\right)^2\right],$$

$$\mathbf{x} = (x_1, x_2, \ldots, x_n), \delta = (\delta_1, \delta_2, \ldots, \delta_n).$$

Show that if $J_\delta u = (\rho_\delta * u)(\mathbf{x}) = \int_{\mathbb{R}^n} \rho_\delta(\mathbf{x} - \mathbf{s})u(\mathbf{s})\,d\mathbf{s}$ and $\|\nabla u\|_{\infty, K_{\hat{\delta}}} \leq M$, then

$$\|J_\delta u - u\|_{\infty, K_{\hat{\delta}}} \leq 4nM\hat{\delta}, \qquad \hat{\delta} = \max_{i \leq i \leq n} (\delta_i) \leq \text{dist}(K_{\hat{\delta}}, \partial I),$$

with $K_{\hat{\delta}}$ a suitable compact subset of

$$I = \overbrace{[0, 1] \times [0, 1] \times \cdots \times [0, 1]}^{n \text{ times}} \subset \mathbb{R}^n.$$

6.4. With the same notation as in Exercise 6.3, show that if $u, u_m \in C^0(I)$ (continuous functions over I) and $\|u_m - u\|_{\infty, I} \leq \varepsilon$, then $\|\nabla J_\delta u_m - \nabla J_\delta u\|_{\infty, K_{\hat{\delta}}} \leq 8\varepsilon/\tilde{\delta}$, where $\tilde{\delta} = \min(\delta_1, \delta_2, \ldots, \delta_n)$.

6.13   REFERENCES AND COMMENTS

The following references and comments serve to expand the basic material covered in the corresponding sections.

6.1–6.3. Here we follow in part the article

D. A. Murio and D. Hinestroza, Numerical identification of forcing terms by discrete mollification, *Comput. Math. Appl.* **17** (1990), 75–89.

6.4. Several algorithms have been proposed for the solution of the parameter identification problem in the steady-state flow. An important review of parameter estimation for ground water hydrology was presented in

W. W. G. Yeh, Review of parameter identification procedures in ground water hydrology: the inverse problem, *Water Resources Res.* **22** (1986), 95–108.

The output least squares minimization method was first considered by

E. Frind and G. Pinder, Galerkin solutions of the inverse problem for aquifer transmissivity, *Water Resources Res.* **9** (1973), 1397–1410.

The method has been recently studied by

K. Kunish and L. W. White, Identifiability under approximation for an elliptic boundary value problem, *SIAM J. Control Optim.* **25** (1987), pp. 279–297.

and, combined with regularization techniques, by

C. Kravaris and J. H. Seinfeld, Identification of parameters in distributed parameter systems by regularization, *SIAM J. Control Optim.* **23** (1985), 217–241.

Stability results have been given by

R. S. Falk, Error estimates for the numerical identification of a variable coefficient, *Math. Comp.* **40** (1983), 537–546.

Further applications of this method to different parameter identification problems can be found in

G. Chavent, Identification of distributed parameter systems: about the output least squares method, its implementation and identifiability, *Proc. Fifth IFAC Symp. on Identification and System Parameter Estimation* **1**, R. Iserman (ed.), Pergamon, Darnstadt, 1980, pp. 85–97.

H. T. Banks and K. Kunish, Parameter estimation techniques for nonlinear distributed parameter systems, *Nonlinear Phenomena in Mathematical Sciences*, V. Lakshmikantham (ed.), Academic, New York, 1982, pp. 57–67.

A new interesting variational approach based on the minimization of a convex functional was introduced in

R. V. Khon and B. D. Lowe, A variational method for parameter estimation, *Math. Modeling Numer. Anal.* **22** (1988), 119–158.

Singular perturbation techniques have been proposed in

G. Alessandrini, An identification problem for an elliptic equation in two variables, *Ann. Mat. Pura Appl.* **145** (1986), 265–296.

G. Alessandrini, On the identification of the leading coefficient of an elliptic equation. *Boll. U.M.I., Analisi Funzionale Applicazioni, Ser. VI* **4-C** (1985), 87–111.

Finite-difference "marching" methods, integrating along the lines of steepest descent or characteristics, were first developed—*for noiseless data*—by

G. R. Richter, An inverse problem for the steady state diffusion equation, *SIAM J. Appl. Math.* **41** (1981), 210–221.

G. R. Richter, Numerical identification of a spatially varying diffusion coefficient, *Math. Comp.* **36** (1981), 404–427.

6.5–6.7. In these sections we follow closely the treatment in

D. Hinestroza and D. A. Murio, Identification of transmissivity coefficients by mollification techniques. I. One-dimensional elliptic and parabolic problems, *Comput. Math. Appl.* **25** (1993), 59–79.

6.8. The most commonly used method for the approximate determination of unknown coefficients in the one-dimensional transient flow case is the output least squares approach indicated in the previous references. Some alternative approaches including asymptotic imbedding and adaptive control techniques have been investigated in

H. W. Alt, K. H. Hoffmann, and J. Sprekel, A numerical procedure to solve certain identification problems, *Internat. Ser. Numer. Math.* **68** (1984), 11–43.

J. Baumeister and W. Scondo, Asymptotic imbedding methods for parameter estimation, *Proc. 26th. Conf. on Decision and Control, Los Angeles,* 1987, pp. 170–174.

For spatially varying coefficients, a finite-difference scheme is discussed in

R. E. Ewing and T. Lin, Parameter identification problems in single-phase and two phase flow, *Internat. Ser. Numer. Math.* **91** (1989), 85–108.

A new method that combines the output least squares and the error equation, resulting in an optimization problem solved by an augmented Lagrangian approach, can be found in

K. Kunish and G. Peichel, Estimation of temporally and spatially varying diffusion coefficient in a parabolic system by an augmented Lagrangian technique, *Numer. Math.* **59** (1991), 473–509.

6.9–6.11. Here we follow the article of D. Hinestroza and D. A. Murio referenced previously.

For the application of the mollification method to the two-dimensional identification of transmissivity coefficients, see

D. Hinestroza, Identification of transmissivity coefficients by mollification techniques. II. Two-dimensional elliptic and parabolic problems, to appear.

APPENDIX A

MATHEMATICAL BACKGROUND

There are several basic concepts concerning properties of Hilbert spaces, mollifiers, Fourier transforms, and discrete functions that play an important role in the topics developed in this appendix.

We state all the necessary definitions and propositions but offer only a selected set of demonstrations. For the remaining proofs, the interested reader should consult the references indicated at the end of the chapter.

Throughout this appendix, the space of real numbers (the real line) is denoted $\mathbb{R}$, the complex numbers $\mathbb{C}$, and $\mathbb{R}^n = \mathbb{R} \times \mathbb{R} \times \cdots \times \mathbb{R}$ (n times) denotes the n-dimensional Euclidean space. The length of the vector $x = (x_1, x_2, \ldots, x_n) \in \mathbb{R}^n$ is represented by

$$\|x\| = \left\{ \sum_{i=1}^{n} x_i^2 \right\}^{1/2}$$

and Ω always denotes an open, bounded domain in $\mathbb{R}^n$.

In the sequel additional notation and conventions will be defined where they first appear.

A.1 L^p SPACES

DEFINITION A.1 $L^p(\Omega)$, $p \geq 1$, denotes the space of all equivalence classes of real-valued (or complex-valued) Lebesgue-measurable functions f, defined almost everywhere on an open, bounded domain $\Omega \subset \mathbb{R}^n$, whose

pth powers $|f|^p$ are Lebesgue integrable over Ω, that is,

$$\int_\Omega |f(x)|^p\, dx < \infty,$$

$x = (x_1, x_2, \ldots, x_n) \in \mathbb{R}^n$ and $dx = \mathrm{mes}(dx_1, dx_2, \ldots, dx_n)$.

Any pointwise property attributed to a function $f \in L^p(\Omega)$ will thus be understood to hold in the usual sense for some function in the same equivalence class. We recall that the function f is equivalent to the function g, if $f(x) = g(x)$ for almost all $x \in \Omega$, that is, except for a set of n-dimensional measure 0.

We notice that if $f(x)$ and $g(x)$ belong to $L^p(\Omega)$, using the inequalities

$$|f(x) + g(x)|^p \le 2^p \max\{|f(x)|^p, |g(x)|^p\} \le 2^p\{|f(x)|^p + |g(x)|^p\},$$

it follows that

$$(f + g)(x) \equiv f(x) + g(x) \in L^p(\Omega)$$

and

$$(cf)(x) \equiv cf(x) \in L^p(\Omega) \qquad \text{for every real (complex) number } c.$$

The set of functions $L^p(\Omega)$ with the addition and scalar multiplication defined previously is a vector space.

DEFINITION A.2   The real function

$$\|f\|_{p,\Omega} = \left(\int_\Omega |f(x)|^p\, dx \right)^{1/p}, \qquad \forall f \in L^p(\Omega),$$

is a norm in the vector space $L^p(\Omega)$; that is, the following axioms are satisfied:

(N1) $\|f\|_{p,\Omega} \ge 0$, $\forall f \in L^p(\Omega)$,

(N2) $\|f\|_{p,\Omega} = 0$ if and only if $f = 0$,

(N3) $\|cf\|_{p,\Omega} = |c|\, \|f\|_{p,\Omega}$, $\forall c$ scalar, $\forall f \in L^p(\Omega)$,

(N4) $\|f + g\|_{p,\Omega} \le \|f\|_{p,\Omega} + \|g\|_{p,\Omega}$, $\forall f, g \in L^p(\Omega)$.

From property (N4), the triangle inequality, we find

(N4') $\|f - g\|_{p,\Omega} \ge |\,\|f\|_{p,\Omega} - \|g\|_{p,\Omega}\,|$, $\forall f, g \in L^p(\Omega)$.

Proof

$$\|f\|_{p,\Omega} = \|(f - g) + g\|_{p,\Omega} \leq \|f - g\|_{p,\Omega} + \|g\|_{p,\Omega}$$

and

$$\|f\|_{p,\Omega} - \|g\|_{p,\Omega} \leq \|f - g\|_{p,\Omega}.$$

Similarly,

$$\|g\|_{p,\Omega} - \|f\|_{p,\Omega} \leq \|g - f\|_{p,\Omega} = \|(-1)(f - g)\|_{p,\Omega} = \|f - g\|_{p,\Omega}. \quad \blacksquare$$

We shall use the notation $\|f\|_p$ for $\|f\|_{p,\Omega}$ and $\|f\|$ for $\|f\|_2$ when there is no ambiguity. $(L^p(\Omega), \|\cdot\|_p)$—briefly $L^p(\Omega)$—is a normed space. The distance, as usual, is defined by the real function (functional)

$$d(f, g) = \|f - g\|_{p,\Omega},$$

which satisfies properties

(d1) $d(f, g) \geq 0$, $\forall f, g \in L^p(\Omega)$,

(d2) $d(f, g) = 0$ if and only if $f = g$,

(d3) $d(f, g) = d(g, f)$, $\forall f, g \in L^p(\Omega)$,

(d4) $d(f, h) \leq d(f, g) + d(g, h)$, $\forall f, g, h \in L^p(\Omega)$.

Theorem A.1 (Riesz and Fischer)   The normed space $L^p(\Omega)$ is complete.

In other words, $L^p(\Omega)$ is a Banach space. This means that for every sequence $\{f_n\}_{n=1}^{\infty}$ of functions in $L^p(\Omega)$ such that $\|f_n - f_m\|_p \to_{n,m \to \infty} 0$, there exists $f \in L^p(\Omega)$ such that $\|f_n - f\|_p \to_{n \to \infty} 0$.

For $p = \infty$, $L^{\infty}(\Omega)$ denotes the Banach space of bounded functions on Ω with the norm $\|f\|_{\infty,\Omega} = \sup_{x \in \Omega}|f(x)|$. In this context, the supremum (infimum) of the function f should be understood as the essential supremum (infimum).

Note: The minimum essential upper bound σ of $f(x)$ is called the essential supremum of $f(x)$; σ is an essential upper bound of $f(x)$ if $f(x) \leq \sigma$ for almost every x, that is, there exists a set P of measure 0 $(P \subset \Omega)$ such that $f(x) \leq \sigma$ if $x \notin P$.

Theorem A.2 (Hölder)   If $1/p + 1/q = 1$, $p \geq 1$, $q \geq 1$ and if $f \in L^p(\Omega)$, $g \in L^q(\Omega)$, then fg is integrable, that is, $fg \in L^1(\Omega)$ and

$$\left| \int_{\Omega} f(x) g(x)\, dx \right| \leq \|f\|_p \|g\|_q.$$

Observe that for $p = 2$, we have $q = 2$ and Hölder's inequality reduces to the Cauchy–Schwarz inequality. (See Theorem A.3.)

EXERCISE A.1

1. Show that $L^p(\Omega) \subset L^1(\Omega)$, $p \geq 1$.

2. Prove Minkowski's inequality

$$\int_\Omega |f(x) + g(x)|^p \, dx \leq \left\{ \int_\Omega |f(x)|^p \, dx \right\}^{1/p} + \left\{ \int_\Omega |g(x)|^p \, dx \right\}^{1/p},$$

$$\forall f, g \in L^p(\Omega).$$

3. Show that if $f \in L^p(\Omega)$ and $g \in L^q(\Omega)$, where $1/p + 1/q = 1$, then $fg \in L^1(\Omega)$.

4. Show that $L^p(\Omega)$ is a metric space—satisfies (d1)–(d4)—with

$$d(f, g) = \|f - g\|_p, \quad \forall f, g \in L^p(\Omega).$$

5. For what values of p does the function $f(x) = x^{-1/2}$ belong to $L^p((0, 1))$?

6. Prove that if $f_n \to_{n \to \infty} f$ in $L^p(\Omega)$ and $g \in L^q(\Omega)$ (p and q as in Exercise 3), then $\lim_{n \to \infty} \int_\Omega f_n(x) g(x) \, dx = \int_\Omega f(x) g(x) \, dx$.

A.2   THE HILBERT SPACE $L^2(\Omega)$

We observe that if $f \in L^p(\Omega)$ and $g \in L^p(\Omega)$ with $p \neq 2$, $1 \leq p \leq \infty$, Hölder's theorem is not applicable, that is, in general fg will not be integrable. However, for $p = 2$, it is possible to introduce a geometrical structure: the bilinear form

$$(f, g) = \int_\Omega f(x) \overline{g(x)} \, dx, \quad \forall f, g \in L^2(\Omega), \tag{A.1}$$

defines an inner product in $L^2(\Omega)$, that is, it verifies

(H1) $(f, f) = \|f\|^2$, $\forall f \in L^2(\Omega)$,

(H2) $(f, g) = \overline{(g, f)}$, $\forall f, g \in L^2(\Omega)$,

(H3) $(cg, f) = c(f, g)$, $\forall c \in \mathbb{C}, \forall f, g \in L^2(\Omega)$,

(H4) $(f + g, h) = (f, h) + (g + h)$, $\forall f, g, h \in L^2(\Omega)$.

Here, $\overline{(g, f)}$ stands for the complex conjugate of the complex number (g, f).

From these properties it follows that

(H5) $(f, cg) = \bar{c}(f, g), \qquad \forall\, c \in \mathbb{C}, \forall\, f, g \in L^2(\Omega),$

and

(H6) $(f, g + h) = (f, g) + (f, h), \qquad \forall\, f, g, h \in L^2(\Omega).$

Proof

$$(f, cg) = \overline{(cg, f)} = \bar{c}\,\overline{(g, f)} = \bar{c}(f, g).$$

Also,

$$(f, g + h) = \overline{(g + h, f)} = \overline{(g, f)} + \overline{(h, f)} = (f, g) + (f, h). \quad \blacksquare$$

DEFINITION A.3   A Hilbert space H is a Banach space—complete normed vector space—with an inner product $(\cdot\,, \cdot\,)$ that verifies properties (H1) to (H4).
Note: The space $L^2(\Omega)$ with the inner product (A.1) is a Hilbert space.

DEFINITION A.4   Two functions f and g in $L^2(\Omega)$ are said to be orthogonal, and we write $f \perp g$ if $(f, g) = 0$.

Theorem A.3 (Cauchy–Schwarz Inequality)   In a Hilbert space H,

$$|(\mathbf{u}, \mathbf{v})| \le \|\mathbf{u}\|\,\|\mathbf{v}\|, \qquad \forall\, \mathbf{u}, \mathbf{v} \in H.$$

Proof. Consider $\mathbf{v} \ne \mathbf{0}$. From

$$\begin{aligned}
0 \le \|\mathbf{u} - c\mathbf{v}\|^2 &= (\mathbf{u} - c\mathbf{v}, \mathbf{u} - c\mathbf{v}) \\
&= (\mathbf{u}, \mathbf{u}) + (\mathbf{u}, -c\mathbf{v}) + (-c\mathbf{v}, \mathbf{u}) + (-c\mathbf{v}, -c\mathbf{v}) \\
&= \|\mathbf{u}\|^2 - \bar{c}(\mathbf{u}, \mathbf{v}) - c\,\overline{(\mathbf{u}, \mathbf{v})} + |c|^2\|\mathbf{v}\|^2,
\end{aligned}$$

with $c = (\mathbf{u}, \mathbf{v})/\|\mathbf{v}\|^2$, we get $|(\mathbf{u}, \mathbf{v})|^2 \le \|\mathbf{u}\|^2\|\mathbf{v}\|^2.$ $\blacksquare$

We observe that the inner product with properties (H1) to (H4) determines the norm in the space.
Again, if $p \ne 2$ we cannot talk about orthogonality or angles in $L^p(\Omega)$ and we can only consider distances. $L^2(\Omega)$ is the only $L^p(\Omega)$ space that is a Hilbert space.
A function f that belongs to $L^2(\Omega)$ is said to be square integrable in Ω.
Notice that if a function $f(x)$ is continuous in $\bar{\Omega}$ (the closure of Ω), then from $\int_\Omega f(x)\,dx = 0$ it follows that $f(x) \equiv 0$ in $\bar{\Omega}$. If $f \in L^2(\Omega)$, without the

continuity assumption, our conclusion should be that $f(x) = 0$ almost every-where, that is, except perhaps on a set of measure 0. For instance, the function

$$f(x) = \begin{cases} 0, & 0 \le x < 1/2, \\ 1, & x = 1/2, \\ 0, & 1/2 < x \le 1, \end{cases}$$

is square integrable in $(0, 1)$ but it is not identically 0 there. At $x = 1/2$—or at any other set of measure 0—the function need not even be defined. Two functions $f(x)$ and $g(x)$ which are equivalent in $L^2(\Omega)$ are characterized by the property

$$\int_\Omega [f(x) - g(x)]^2\, dx = 0.$$

EXERCISE A.2

1. Show that if $f \in L^2(\Omega)$ is orthogonal to all the elements of a subset D which is dense in $L^2(\Omega)$, then $f = 0$.

2. Let $f_n(x) = n/(1 + nx^{1/2})$ for $0 \le x \le 1$, $n = 1, 2, \dots$.
   (a) Is the sequence $\{f_n\}_{n=1}^\infty$ a Cauchy sequence?
   (b) Does $f_n \in L^2((0, 1))\ \forall\ n$?
   (c) Does $\lim_{n \to \infty} f_n(x) \in L^2((0, 1))$?

3. Let $\{\beta_j\}_{j=1}^\infty$ be an orthonormal sequence of functions in $L^2((a, b))$, that is, $(\beta_i, \beta_j) = 0\ \forall\ i, j,\ i \ne j$, $\|\beta_j\|_2 = 1\ \forall\ j$. Let $S_n(x) = \sum_{k=1}^n c_k \beta_k(x)$, where $c_k = \int_a^b f(x)\beta_k(x)\, dx$. Show that for any function $f \in L^2((a, b))$,

$$\|S_n - f\|_2 = \int_a^b |f(x)|^2\, dx - \sum_{k=1}^n c_k^2.$$

   (The coefficients c_k are called Fourier coefficients.)

4. Under the conditions of Exercise 3, show that

$$\sum_{k=1}^\infty c_k^2 \le \int_a^b |f(x)|^2\, dx \qquad \text{(Bessel's inequality).}$$

5. Under the conditions of Exercise 3, prove that if $\lim_{n \to \infty} \|S_n - f\|_2 = 0$, then

$$\int_a^b |f(x)|^2\, dx = \sum_{k=1}^\infty c_k^2 \qquad \text{(Parseval's identity).}$$

A.3   APPROXIMATION OF FUNCTIONS IN $L^2(\Omega)$

DEFINITION A.5 $C^\infty(\overline{\Omega})$ denotes the set of all functions which are continuous together with their derivatives of all orders in $\overline{\Omega}$.

DEFINITION A.6   The support of a function $f(x)$ in Ω, written supp f, is the closure in $\mathbb{R}^n$ of the set of points $x \in \overline{\Omega}$ for which $f(x) \neq 0$.

DEFINITION A.7 $C_0^\infty(\Omega)$ is the set of functions from $C^\infty(\overline{\Omega})$ with compact support in Ω.

Observe that if $f \in C_0^\infty(\Omega)$, then $f \in C^\infty(\overline{\Omega})$ and supp $f \subset \Omega$. Moreover, since supp f is a closed set contained in an open domain Ω, there is a positive distance from any point in supp f to the boundary $\delta\Omega$ of the domain Ω. We say, briefly, that $f(x)$ is 0 in a neighborhood of the boundary. For instance, the function

$$f(x) = \begin{cases} \exp[1/(x^2 - 4)], & -2 < x < 2, \\ 0, & \text{otherwise}, \end{cases}$$

is $C^\infty(\mathbb{R})$ and, relative to the domain $\Omega = (-5, 5)$, for example, supp $f = [2, 2]$. Thus, $f \in C_0^\infty(\Omega)$.

With more generality, we can consider the function

$$\varphi_\delta(x) = \begin{cases} \exp[1/(x^2 - \delta^2)], & -\delta < x < \delta, \\ 0, & \text{otherwise}, \end{cases}$$

where δ is a fixed positive number. $\varphi_\delta \in C^\infty(\mathbb{R})$ and supp $\varphi_\delta = [-\delta, \delta]$. If we compute $\int_{-\infty}^\infty \varphi_\delta(x)\,dx = h(\delta)$, the function $\psi_\delta(x) = \varphi_\delta(x)/h(\delta)$ is such that $\int_{-\infty}^\infty \psi_\delta(x)\,dx = 1$.

Now consider an arbitrary given function $f \in L^2([a, b])$, extended to the entire real line by setting $x = 0$ if $x \notin [a, b]$ and introduce the function

$$f_\delta(x) = (f * \psi_\delta)(x) = \int_{-\infty}^\infty f(s)\psi_\delta(x - s)\,ds = \int_{-\infty}^\infty f(x - s)\psi_\delta(s)\,ds.$$

Notice that f_δ, the convolution of the functions f and ψ_δ, is such that $f_\delta \in C^\infty(\mathbb{R})$ and supp $f_\delta = [a - \delta, b + \delta]$; that is, $f_\delta \in C_0^\infty(\mathbb{R})$. Moreover, $\lim_{\delta \to 0} f_\delta(x) = f(x) \ \forall \ f \in L^2([a, b])$. In effect, we have

$$\|f - f_\delta\|_{L^2(\mathbb{R})}^2 = \int_{-\infty}^\infty |f(x) - f_\delta(x)|^2\,dx$$

$$= \int_{-\infty}^\infty \left| \int_{-\infty}^\infty \{f(x)\psi_\delta(x - s) - f(s)\psi_\delta(x - s)\}\,ds \right|^2 dx.$$

For the inner integral, we set $y = s - x$ and $dy = ds$. Hence,

$$\left| \int_{-\infty}^{\infty} [f(x) - f(s)] \psi_\delta(-y) \, dy \right|^2 = \left| \int_{-\infty}^{\infty} [f(x) - f(x+y)] \psi_\delta(-y) \, dy \right|^2$$

$$= \left| \int_{-\infty}^{\infty} [f(x) - f(s)] \psi_\delta(y) \, dy \right|^2$$

$$= \left| \int_{-\delta}^{\delta} [f(x) - f(x+y)] \psi_\delta(y) \, dy \right|^2$$

$$\le \int_{-\delta}^{\delta} |f(x) - f(x+y)|^2 \, dy \int_{-\delta}^{\delta} |\psi_\delta(y)|^2 \, dy$$

$$\le 2\delta C_1 \int_{-\delta}^{\delta} |f(x) - f(x+y)|^2 \, dy,$$

using the Cauchy–Schwarz inequality and the fact that $\int_{-\delta}^{\delta} |\psi_\delta(y)|^2 \, dy \le C_1$, where C_1 is a constant independent of δ.

Thus, we have

$$\|f - f_\delta\|_{L^2(\mathbb{R})}^2 \le 2\delta C_1 \int_{-\infty}^{\infty} \int_{-\delta}^{\delta} |f(x) - f(x+y)|^2 \, dy \, dx$$

$$= 2\delta C_1 \int_{-\delta}^{\delta} \int_{-\infty}^{\infty} |f(x) - f(x+y)|^2 \, dx \, dy.$$

Now, for every $\varepsilon > 0$, there exists $\gamma(\varepsilon) > 0$ such that, whenever $|y| < \gamma$,

$$\int_{-\infty}^{\infty} |f(x) - f(x+y)|^2 \, dx < \varepsilon^2$$

due to the continuity of functions in $L^2(\Omega)$ with respect to the L^2 norm. [See Smirnov (1964), page 200, for a proof.]

Hence, choosing $0 < \delta < \gamma$,

$$\|f - f_\delta\|_{L^2(\mathbb{R})}^2 \le 2\delta C_1 \varepsilon^2 \int_{-\delta}^{\delta} dy = 4\delta^2 C_1 \varepsilon^2 = C\varepsilon^2.$$

Clearly, we can pick ε so that γ and, consequently, δ are as small as desired. Also notice that

$$\|f - f_\delta\|_{L^2((a,b))}^2 \le \|f - f_\delta\|_{L^2(\mathbb{R})}^2. \quad \blacksquare$$

We have shown the following result.

Theorem A.4 If $f \in L^2([a, b])$, then $f_\delta \in C^\infty([a, b])$, supp $f_\delta = [a - \delta, b + \delta]$, and $\lim_{\delta \to 0} \|f - f_\delta\|_{L^2([a, b])} = 0$.

We are now ready for the main density result.

Theorem A.5 The set $C_0^\infty[a, b]$ is dense in $L^2([a, b])$.

Proof. We need to show that for every function $f \in L^2([a, b])$ and for every $\varepsilon > 0$, there exists a function $g \in C_0^\infty[a, b]$ such that $\|f - g\|_{L^2([a, b])} < \varepsilon$.

We will assume that the following property holds for f: given $\eta > 0$, there exists $\mu > 0$ such that

$$\int_a^{a+\mu} f^2(x)\, dx + \int_{b-\mu}^b f^2(x)\, dx < 2\eta. \tag{A.2}$$

[See Royden (1968), page 85, Proposition 13.]

Let $z(x)$ be a function defined in $[a, b]$ by

$$z(x) = \begin{cases} f(x), & \text{if } a + \mu \le x \le b - \mu, \\ 0, & \text{if } a \le x < a + \mu \text{ or } b - \mu < x \le b. \end{cases}$$

The function $z(x) \in L^2([a, b])$ and is a cutoff version of $f(x)$. It follows, from inequality (A.2), that

$$\|f - z\|_{L^2([a, b])}^2 = \int_a^b [f(x) - z(x)]^2\, dx$$

$$= \int_a^{a+\mu} f^2(x)\, dx + \int_{b-\mu}^b f^2(x)\, dx < 2\eta.$$

If we set $\eta = \varepsilon^2/8$,

$$\|f - z\|_{L^2([a, b])} \le \frac{\varepsilon}{2}. \tag{A.3}$$

Now pick $\delta_1 > 0$ and construct the function $z_{\delta_1} \in C^\infty(\mathbb{R})$. Notice that since $z(x)$ is identically 0 in $[a, a + \delta_1] \cup (b - \delta_1, b]$, the function $z_{\delta_1}(x)$ has compact support in (a, b) for every $\delta_1 < \mu$.

According to Theorem A.4, given $\varepsilon/2$, it is possible to find $\delta_2 > 0$ such that $\|z - z_{\delta_2}\|_{L^2([a, b])} \le \varepsilon/2$.

Let $\delta = \min(\delta_1, \delta_2)$. Then the function $z_\delta \in C_0^\infty(a, b)$ and

$$\|z - z_\delta\|_{L^2([a, b])} \le \frac{\varepsilon}{2}. \tag{A.4}$$

Consequently, from (A.3) and (A.4),

$$\|f - z_\delta\|_{L^2([a,b])} = \|f - z + z - z_\delta\|_{L^2([a,b])}$$
$$\leq \|f - z\|_{L^2([a,b])} + \|z - z_\delta\|_{L^2([a,b])}$$
$$\leq \frac{\varepsilon}{2} + \frac{\varepsilon}{2} = \varepsilon. \quad \blacksquare$$

Finally, we observe that Theorems A.4 and A.5 can be extended to $\mathbb{R}^n$, covering the case of functions of several variables in quite similar manner.

EXERCISE A.3

1. If $f(x) \in L^1([a,b])$, show that $\lim_{h \to 0} \int_a^b |f(x+h) - f(x)| \, dx = 0$, assuming that given $\varepsilon > 0$, there exists a continuous function $g(x)$ such that $\int_a^b |f(x) - g(x)| \, dx < \varepsilon$.

2. If $f \in L^1(\mathbb{R}^n)$ and $g \in L^p(\mathbb{R}^n)$, $1 \leq p < \infty$, prove that $f * g \in L^p(\mathbb{R}^n)$ and $\|f * g\|_p \leq \|f\|_1 \|g\|_p$.

3. Show that if $f, g, h \in L^1(\mathbb{R}^n)$, then

   (a) $\qquad\qquad f * g = g * f \qquad\qquad$ (Commutative law),

   (b) $\qquad f * (g + h) = f * g + f * h \qquad$ (Distributive law),

   (c) $\qquad (f * g) * h = f * (g * h) \qquad$ (Associative law).

A.4   MOLLIFIERS

DEFINITION A.8   A function $f: \mathbb{R}^n \to \mathbb{R}$ is locally integrable if for each compact set $K \subseteq \mathbb{R}^n$, $f \in L^1(K)$.

In order to treat local properties and generalize the ideas utilized in the last theorems of Sec. A.3, we introduce the concept of mollification.

Consider $\Omega \subset \mathbb{R}^n$ and let x be an arbitrary point in $\mathbb{R}^n$. Define a function $w(x)$ that satisfies

(i) $w \in C^\infty(\mathbb{R}^n)$.

(ii) $w(x) \equiv 0$ for $\|x\| \geq 1$ and $w(x) \geq 0$ for $\|x\| < 1$.

(iii) $\int_{\mathbb{R}^n} w(x) \, dx = 1$.

The function $w(x)$ is nonnegative, has total integral 1, has continuous derivatives of all orders on $\mathbb{R}^n$, and vanishes (together with all its partial derivatives) outside $B(0,1) = \{x \in \mathbb{R}^n: \|x\| < 1\}$ (the unit ball centered at 0 and radius 1). We notice that $w \in C_0^\infty(B)$.

For example, w can be the function

$$w(x) = \begin{cases} C \exp\left[-\dfrac{1}{1 - \|x\|^2}\right], & \text{for} \quad \|x\| < 1, \\ 0, & \text{for} \quad \|x\| \geq 1, \end{cases}$$

where C is a suitable constant chosen such that w satisfies (iii).

Next, for any real number $\delta > 0$, we define the family of functions

$$w_\delta(x) = \frac{1}{\delta^n} w\left(\frac{x}{\delta}\right)$$

and observe that $w_\delta \in C_0^\infty(B(0, \delta))$, where $B(0, \delta)$ indicates the ball centered at 0 and radius δ. Moreover, $\int_\Omega w_\delta(x)\, dx = 1$.

DEFINITION A.9 For any locally integrable function f in Ω, the mollifier J_δ is defined by

$$J_\delta f(x) = \int_{\mathbb{R}^n} w_\delta(x - y) f(y)\, dy.$$

The function $J_\delta f(x)$ is defined at all points $x \in \Omega$ such that $\text{dist}(x, \delta\Omega) = \inf_{z \in \partial\Omega} \|x - z\| > \delta$.

Notice that if $\text{supp } f \subseteq K$, with K a compact subset of Ω, the choice $\delta < \text{dist}(K, \partial\Omega)$ implies $\text{supp } J_\delta f \subset \Omega$, that is, $J_\delta f \in C_0^\infty(\Omega)$. (See the construction of z_δ in the proof of Theorem A.5.)

Remark. If $f \in L^p(\mathbb{R}^n)$,

$$J_\delta f(x) = \int_{\mathbb{R}^n} w_\delta(x - y) f(y)\, dy = \int_{\mathbb{R}^n} w_\delta(y) f(x - y)\, dy$$

is the convolution $(w_\delta * f)(x)$ and $\text{supp } J_\delta f \subseteq \text{supp } w_\delta + \text{supp } f$ since $(w_\delta * f)(x) \neq 0$ implies $x - y \in \text{supp } f$ for some $y \in \text{supp } w_\delta$. ∎

The importance of $J_\delta f$ arises from the fact that $J_\delta f$ behaves much like f, but it is very smooth. The number $J_\delta f(x)$ can be interpreted as a w_δ-weighted average of the values of the function f in the ball $\|x\| < \delta$. The kernel w_δ is said to "scan" across the function f. For very small values of δ, $J_\delta f$ is a uniformly close approximation to f, provided that f is continuous. More precisely, we have the following result.

Theorem A.6 Let f be a locally integrable function on $\Omega \subseteq \mathbb{R}^n$. Then $J_\delta f \in C^\infty(\Omega)$.

Proof. From the continuity of $w_\delta(x)$, given $\varepsilon > 0$, there exists $\eta(\varepsilon) > 0$ such that if $\|\Delta x\| < \eta$, then $|w_\delta(x + \Delta x - y) - w_\delta(x - y)| < \varepsilon$. Thus, if $\|\Delta x\| < \eta$,

$$|J_\delta f(x + \Delta x) - J_\delta f(x)| = \left| \int_\Omega f(y)[w_\delta(x + \Delta x - y) - w_\delta(x - y)]\, dy \right|$$

$$\leq \varepsilon \left| \int_\Omega f(y)\, dy \right|.$$

Since $f(x)$ is, by hypothesis, integrable, the last integral is finite and the continuity of $J_\delta f(x)$ follows.

From the definition of $J_\delta f$, the differentiation can be carried under the integral sign, so that the differentiability properties of $J_\delta f$ follow from those of w_δ. ■

Theorem A.7   If $f \in L^p(\Omega)$, then $\|J_\delta f\|_{p,\Omega} \leq \|f\|_{p,\Omega}$.

Proof. By Hölder's inequality, with $1/p + 1/q = 1$, we have

$$|J_\delta f(x)|^p = \left| \int_\Omega f(y)[w_\delta(x - y)]^{1/p}[w_\delta(x - y)]^{1/q}\, dy \right|^p$$

$$\leq \int_\Omega |f(y)|^p |w_\delta(x - y)|\, dy \left[\int_\Omega |w_\delta(x - y)|\, dy \right]^{p/q}$$

$$= \int_\Omega |f(y)|^p |w_\delta(x - y)|\, dy.$$

Integrating and using Fubini's theorem,

$$\|J_\delta f\|_{p,\Omega}^p = \int_\Omega |J_\delta f(x)|^p\, dx \leq \int_\Omega \left[\int_\Omega |f(y)|^p |w_\delta(x - y)|\, dy \right] dx$$

$$= \int_\Omega |f(y)|^p \left[\int_\Omega |w_\delta(x - y)|\, dx \right] dy = \int_\Omega |f(y)|^p\, dy = \|f\|_{p,\Omega}^p. ■$$

Theorem A.8   If f is continuous at a point $x \in \mathbb{R}^n$, then $\lim_{\delta \to 0} |J_\delta f(x) - f(x)| = 0$, the convergence being uniform on any compact set of continuity points.

Proof. Extending f to $\mathbb{R}^n$ by setting $f \equiv 0$ outside Ω, we have

$$|J_\delta f(x) - f(x)| \leq \int_{\mathbb{R}^n} w_\delta(x - y)|f(y) - f(x)|\, dy$$

$$\leq \sup_{\|y - x\| < \delta} |f(y) - f(x)|.$$

The right-hand side tends to 0 with δ at each point x and the convergence to 0 is uniform for any compact set of continuity points.   ■

Obvious modifications of the proof of Theorem A.4 give the more general result.

Theorem A.9   If $f \in L^p(\Omega)$, then $\lim_{\delta \to 0} \|J_\delta f - f\|_{p,\Omega} = 0$.

Proof. The proof is left as an exercise.

EXERCISE A.4

1. Given the function

$$
w(x) = \begin{cases} C \exp\left[-\dfrac{1}{1-x^2}\right], & \text{if} \quad |x| < 1, \\ 0, & \text{if} \quad |x| \geq 1, \end{cases}
$$

   use a simple quadrature formula to approximately determine the constant C such that $\int_{\mathbb{R}} w(x)\, dx = 1$.

2. With the value of C obtained in Exercise 2, use your quadrature formula to approximately evaluate $J_\delta f(x)$ (Definition A.9) if $f(x) = 1$ for $|x| < 1$ and 0 otherwise. Use $\delta = 0.1$, 0.01, and 0.001. In each case, estimate $\|J_\delta f - f\|$, $\|J_\delta f - f\|_\infty$ and check if Theorem A.7 is verified.

3. Prove Theorem A.9.

A.5   FOURIER TRANSFORM

The Fourier transform of an arbitrary square-integrable function allows for a new representation of the function in terms of its "frequency" components. This new characterization is essential for an alternative analysis of the stability of linear models.

DEFINITION A.10   For any function $f \in L^2(\mathbb{R}^n)$, its Fourier transform $\hat{f}$ is defined by

$$
\hat{f}(w) = \frac{1}{(2\pi)^{n/2}} \int_{\mathbb{R}^n} f(x) e^{-ixw}\, dx,
$$

with $x = (x_1, x_2, \ldots, x_n)$, $w = (w_1, w_2, \ldots, w_n)$ points in $\mathbb{R}^n$, $i = \sqrt{-1}$, and $xw = \sum_{j=1}^n x_j w_j$. We observe that $\hat{f}(w)$ is generally a complex function.

DEFINITION A.11   The inverse Fourier transform of $\hat{f}(w)$ is given by

$$f(x) = \left(\hat{f}\right)^{\vee}(x) = \frac{1}{(2\pi)^{n/2}} \int_{\mathbb{R}^n} \hat{f}(w) e^{ixw} \, dw.$$

The L^2 norms of f and its Fourier transform $\hat{f}$ are related as follows.

Theorem A.10 (Parseval's Equality)   For $f \in L^2(\mathbb{R}^n)$, $\|f\|_{2,\mathbb{R}^n} = \|\hat{f}\|_{2,\mathbb{R}^n}$.

Proof. For any functions f and $g \in L^2(\mathbb{R}^n)$,

$$(f,g) = \int_{\mathbb{R}^n} f(x) \overline{g(x)} \, dx = \int_{\mathbb{R}^n} \left[\frac{1}{(2\pi)^{n/2}} \int_{\mathbb{R}^n} \hat{f}(w) e^{ixw} \, dw \right] \overline{g(x)} \, dx$$

$$= \int_{\mathbb{R}^n} \hat{f}(w) \left[\frac{1}{(2\pi)^{n/2}} \int_{\mathbb{R}^n} e^{ixw} \overline{g(x)} \, dx \right] dw \qquad \text{by Fubini's theorem.}$$

Thus,

$$(f,g) = \int_{\mathbb{R}^n} \hat{f}(w) \overline{\left[\frac{1}{(2\pi)^{n/2}} \int_{\mathbb{R}^n} e^{-ixw} g(x) \, dx \right]} \, dw$$

$$= \int_{\mathbb{R}^n} \hat{f}(w) \overline{\hat{g}(w)} \, dw = (\hat{f}, \hat{g}).$$

Setting $f = g$ and taking square roots, we have

$$\|f\|_{2,\mathbb{R}^n} = \|\hat{f}\|_{2,\mathbb{R}^n}. \qquad \blacksquare$$

Note: The Fourier transform preserves inner products and norms in $L^2(\mathbb{R}^n)$. We say that the Fourier transform is an isometry on $L^2(\mathbb{R}^n)$.

Theorem A.11 (Convolution)   For $f, g \in L^2(\mathbb{R}^n)$,

$$\widehat{(f * g)}(w) = (2\pi)^{n/2} \hat{f}(w) \hat{g}(w).$$

Proof

$$\widehat{(f * g)}(w) = \frac{1}{(2\pi)^{n/2}} \int_{\mathbb{R}^n} e^{-ixw} \left[\int_{\mathbb{R}^n} f(s) g(x-s) \, ds \right] dx$$

$$= \int_{\mathbb{R}^n} f(s) \left[\frac{1}{(2\pi)^{n/2}} \int_{\mathbb{R}^n} e^{-ixw} g(x-s) \, dx \right] ds$$

$$= \int_{\mathbb{R}^n} f(s) \left[\frac{1}{(2\pi)^{n/2}} \int_{\mathbb{R}^n} e^{-i(t+s)w} g(t) \, dt \right] ds$$

$$= \int_{\mathbb{R}^n} f(s) \left[e^{-isw} \hat{g}(w) \right] ds = (2\pi)^{n/2} \hat{f}(w) \hat{g}(w). \qquad \blacksquare$$

A similar proof shows the following symmetric property.

Theorem A.12   For $f, g \in L^2(\mathbb{R}^n)$, $(\hat{f} * \hat{g})^{\vee}(x) = (2\pi)^{n/2} f(x) g(x)$.

To investigate the important relationship between Fourier transforms and differentiation of functions in $L^2(\mathbb{R}^n)$, it is convenient to adopt the following notation.

Given a point $x = (x_1, x_2, \ldots, x_n) \in \mathbb{R}^n$ and an ordered n-tuple of non-negative integers $\alpha = (\alpha_1, \alpha_2, \ldots, \alpha_n)$,

$$|\alpha| = \alpha_1 + \alpha_2 + \cdots + \alpha_n,$$
$$x^{\alpha} = x_1^{\alpha_1} \, x_2^{\alpha_2} \quad \cdots \quad x_n^{\alpha_n},$$

and

$$D^{\alpha} f(x) = \frac{\partial^{|\alpha|} f(x)}{\partial_{x_1}^{\alpha_1} \partial_{x_2}^{\alpha_2} \cdots \partial_{x_n}^{\alpha_n}} = \left(\frac{\partial^{\alpha_1}}{\partial_{x_1}^{\alpha_1}} \right) \cdots \left(\frac{\partial^{\alpha_n}}{\partial_{x_n}^{\alpha_n}} \right) f(x).$$

In this manner it is possible to calculate in $\mathbb{R}^n$ in much the same way as in $\mathbb{R}$. For example, we have the following result.

Theorem A.13   If $D^{\alpha} f(x) \in L^2(\mathbb{R}^n)$, then $[\widehat{D^{\alpha} f}](w) = (iw)^{\alpha} \hat{f}(w)$.

Proof

$$f(x) = \frac{1}{(2\pi)^{n/2}} \int_{\mathbb{R}^n} e^{ixw} \hat{f}(w) \, dw,$$

and the following integrals exist by hypothesis

$$D^{\alpha} f(x) = \frac{1}{(2\pi)^{n/2}} \int_{\mathbb{R}^n} D^{\alpha} \left(e^{ixw} \hat{f}(w) \right) dw$$

$$= \frac{1}{(2\pi)^{n/2}} \int_{\mathbb{R}^n} e^{ixw} (iw)^{\alpha} \hat{f}(w) \, dw.$$

Thus,

$$[\widehat{D^{\alpha} f}](w) = (iw)^{\alpha} \hat{f}(w). \quad \blacksquare$$

Remark (Rapid Decay)   Notice that if the Fourier transform of a derivative of f exists, then $\lim_{\|w\| \to \infty} (iw)^{\alpha} \hat{f}(w) = 0$; otherwise the functions will not be in $L^2(\mathbb{R}^n)$, that is, the smoother the function f, the more rapidly the high-frequency components of $\hat{f}$ must approach 0.

The next property relates the support of functions in the real domain with the support of their Fourier transforms.

Theorem A.14 If $f \in L^2(\mathbb{R})$ has compact support, then its Fourier transform cannot equal 0 for $|w| > |w_0| > 0$.

Proof. Without loss of generality, we consider the restriction of an arbitrary function f in $L^2(\mathbb{R})$ to the interval $[-x_0, x_0]$, denoted f_{x_0}. This function can be obtained by multiplying f with the rectangular step function

$$s_{x_0} = \begin{cases} 1, & \text{if} \quad |x| \le x_0, \\ 0, & \text{if} \quad |x| > x_0. \end{cases}$$

Thus,

$$\hat{f}_{x_0}(w) = \left(\widehat{f \cdot s_{x_0}}\right)(w) = \frac{1}{(2\pi)^{n/2}} \left(\hat{f} * \hat{s}_{x_0}\right)(w)$$

according to Theorem A.11.

If we assume that $\hat{f}_{x_0}(w) = 0$ for $|w| > |w_0|$, then

$$\hat{f}_{x_0}(w) = \frac{1}{(2\pi)^{n/2}} \left(\hat{f} * \hat{s}_{x_0}\right)(w) \cdot s_{w_0}(w),$$

where s_{w_0} is the rectangular step function—in the frequency domain—with compact support in $[-w_0, w_0]$.

Taking inverse Fourier transforms in the last expression, we get

$$f_{x_0}(x) = \left(f_{x_0} * \check{s}_{w_0}\right)(x) \qquad \text{for all} \quad x, \ -x_0 \le x \le x_0,$$

which is impossible because $\check{s}_{w_0}(x)$ is not the unit element for the convolution operation in $L^2(\mathbb{R})$. A short calculation shows that

$$\check{s}_{x_0}(x) = \frac{\sqrt{2}}{\sqrt{\pi}} \frac{\sin x w_0}{x}. \qquad \blacksquare$$

EXERCISE A.5

1. If the Fourier transform of $f(x)$ is $\hat{f}(w)$, $x \in \mathbb{R}^n$, show that $\hat{\hat{f}}(-x) = f(w)$.

2. Given the pair $\hat{f}(w) \leftrightarrow f(x)$, $x \in \mathbb{R}^n$, for $a \in \mathbb{R}$, find the Fourier transform of $f(ax)$.

3. Given the pair $\hat{f}(w) \leftrightarrow f(x)$, $x \in \mathbb{R}^n$, for $x_0 \in \mathbb{R}^n$, find the Fourier transform of $f(x - x_0)$.

4. Find the Fourier transform of the Gaussian kernel

$$f(\mathbf{x}) = \frac{1}{\delta\sqrt{\pi}} \exp\left[1 - \frac{\sum_{i=1}^{n} x_i^2}{\delta^2}\right], \qquad \delta \in \mathbb{R}, \delta > 0, x_i \in \mathbb{R}, i = 1, 2, \ldots, n.$$

5. Prove Theorem A.12.

A.6   DISCRETE FUNCTIONS

Consider the real line $\mathbb{R}$ divided into small segments of length (step size) h by the grid points $x_j = jh$, $j \in Z$. (Here Z denotes the set of integer numbers.)
The discrete real (complex) function $f_h = \{f_j\}_{j=-\infty}^{\infty} = \{f(x_j)\}_{j=-\infty}^{\infty} = \{\ldots, f_{-2}, f_{-1}, f_0, f_1, f_2, \ldots\}$ is defined on the grid points x_j, and we usually assume $f_j = 0$ for $|j|$ sufficiently large.

DEFINITION A.12   The inner product of two discrete functions f_h and g_h is defined by

$$(f_h, g_h) = h \sum_{j \in Z} f_j \overline{g_j}.$$

(Note the scaling factor h.)

DEFINITION A.13   The l^2 norm of a discrete function f_h is given by

$$\|f_h\|_2 = (f_h, f_h)^{1/2} = \left\{h \sum_{j \in Z} |f_j|^2\right\}^{1/2}.$$

DEFINITION A.14   The maximum norm of a discrete function f_h is defined as

$$\|f_h\|_\infty = \max_{j \in Z} |f_j|.$$

It is natural to expect that, in the study of finite-difference approximations, the rules for the case of discrete functions may play a similar role as the known rules of calculus for continuous—nondiscrete—functions.

DEFINITION A.15   The forward operator S_+ maps the discrete function $f_h = \{f_j\}_{j=-\infty}^{\infty}$ into the discrete function $S_+ f_h = \{f_{j+1}\}_{j=-\infty}^{\infty}$.

We will also use the operators—in brief notation—

$$(D_+f)_j = \frac{1}{h}(f_{j+1} - f_j) \qquad \text{(Forward difference)},$$

$$(D_-f)_j = \frac{1}{h}(f_j - f_{j-1}) \qquad \text{(Backward difference)},$$

$$(D_0f)_j = \frac{1}{2h}(f_{j+1} - f_{j-1}) \qquad \text{(Centered differences)}.$$

Notice that all these operators can be expressed as linear combinations of "powers" of the forward shift operation S_+. For instance,

$$D_0 = \frac{1}{2h}S_+ - \frac{1}{2h}(S_+)^{-1}, \qquad \text{etc.}$$

DEFINITION A.16 (Discrete Fourier Transform)   Given a discrete function $f_h = \{f_j\}_{j=-\infty}^{\infty}$, its discrete Fourier transform is defined by

$$\tilde{f}(w) = \frac{h}{\sqrt{2\pi}} \sum_{j \in Z} f_j e^{-iwx_j}, \qquad -\frac{\pi}{h} \leq w \leq \frac{\pi}{h}.$$

This is a periodic function in w with period $2\pi/h$ and it can be thought of as a function on the unit circle whose Fourier coefficients are the scaled values of f_h.

The inverse transformation is then given by

$$f_j = \frac{1}{\sqrt{2\pi}} \int_{-\pi/h}^{\pi/h} \tilde{f}(w)e^{iwx_j}\, dw.$$

The relationship between the l^2 norm of the discrete function f_h and the L^2 norm of $\tilde{f}(w)$ is stated as follows.

Theorem A.15

$$\|f_h\|_2 = \|\tilde{f}\|.$$

Proof. Consider a discrete function $f_h = \{f_j\}_{j=-\infty}^{\infty}$, defined at the grid points $x_j = jh$, $j \in Z$.

We begin by introducing Whittaker's cardinal function, defined by

$$W(x) = \frac{\sin(\pi x/h)}{\pi x/h}, \qquad -\infty < x < \infty.$$

Its Fourier transform is given by

$$\hat{W}(x) = \begin{cases} \dfrac{h}{\sqrt{2\pi}}, & -\dfrac{\pi}{h} \leq w \leq \dfrac{\pi}{h}, \\ 0, & \text{otherwise.} \end{cases}$$

In fact,

$$\frac{1}{\sqrt{2\pi}} \int_{-\infty}^{\infty} \hat{W}(w) e^{iwx}\, dw = \frac{h}{2\pi} \int_{-\pi/h}^{\pi/h} e^{iwx}\, dw$$

$$= \frac{h}{\pi} \int_{0}^{\pi/h} \cos wx\, dw = \frac{\sin \pi x / h}{\pi x / h} = W(x).$$

The restriction of $W(x)$ to the grid points gives

$$W(x_j) = \begin{cases} 1, & j = 0, \\ 0, & j \neq 0. \end{cases}$$

We can associate with each point of the grid a shifted cardinal function by defining

$$W_j(x) = \frac{\sin\left[\pi(x - x_j)/h\right]}{\pi(x - x_j)/h}, \qquad -\infty < x < \infty,\, j \in Z.$$

Observe that

$$W_j(x_p) = \begin{cases} 1, & j = p, \\ 0, & j \neq p. \end{cases}$$

Since $W_j(x)$ is obtained by shifting $W(x) = W_0(x)$ by the factor jh, we have $\hat{W}_j(w) = e^{-iwx_j}\hat{W}(w)$.

Also, by Theorem A.10,

$$(W_j, W_p) = \left(\hat{W}_j, \hat{W}_p\right) = \int_{-\infty}^{\infty} \hat{W}_j(w)\, \overline{\hat{W}_p(w)}\, dw = \frac{h^2}{2\pi} \int_{-\pi/h}^{\pi/h} e^{iw(x_j - x_p)}\, dw$$

$$= \frac{h^2}{\pi} \int_{0}^{\pi/h} \cos w(x_j - x_p)\, dw = h\, \frac{\sin\left[\pi(x_j - x_p)/h\right]}{\pi(x_j - x_p)/h}.$$

Thus,

$$(W_j, W_p) = \begin{cases} h, & j = p, \\ 0, & j \neq p, \end{cases}$$

and the sequence of functions $\{W_j\}_{j \in Z}$ is said to be orthogonal.

Now consider the continuous (nondiscrete) function $f^*(x)$ that interpolates the discrete function f_h at the grid points $x_j = jh$, that is, that verifies $f^*(x_j) = f_j$, $j \in Z$. This function can be represented in terms of the sequence of Whittaker's cardinal functions as

$$f^*(x) = \sum_{j \in Z} f_j W_j(x). \qquad (A.5)$$

Applying Fourier transforms to (A.5), we obtain

$$(\widehat{f^*})(w) = \sum_{j \in Z} f_j \hat{W}_j(w) = \sum_{j \in Z} f_j e^{-iwx_j} \hat{W}(w)$$

$$= \frac{h}{\sqrt{2\pi}} \sum_{j \in Z} f_j e^{-iwx_j} = \tilde{f}(w).$$

Hence,

$$(\widehat{f^*})(w) = \begin{cases} \tilde{f}(w), & -\pi/h \le w \le \pi/h, \\ 0, & \text{otherwise,} \end{cases}$$

and by Theorem A.10 it follows that

$$\|f^*\| = \|\tilde{f}\|. \qquad (A.6)$$

On the other hand, taking into account the orthogonality of the sequence of functions $\{W_j\}_{j \in Z}$,

$$\|f^*\| = \left(\sum_{j \in Z} f_j W_j, \sum_{j \in Z} f_j W_j \right)^{1/2} = \left[\sum_{j \in Z} f_j^2 (W_j, W_j) \right]^{1/2}$$

$$= \left[h \sum_{j \in Z} f_j^2 \right]^{1/2} = \|f_h\|_2.$$

From this and equality (A.6), we have $\|\tilde{f}\| = \|f_h\|_2$. ∎

The Fourier transform of a nondiscrete function $f(x)$ and the discrete Fourier transform of the associated discrete function $f_h = \{f_j\}_{j=-\infty}^{\infty}$, obtained by restricting f to the grid points $x_j = jh$, are related by Poisson's summation formula

$$\tilde{f}(w) = \sum_{j=-\infty}^{\infty} \hat{f}\left(w + j\frac{2\pi}{h} \right).$$

[See Papoulis, (1962), page 48, for a proof.]

If the function $f(x)$ is band-limited, that is, $\hat{f}(w) = 0$ for $|w| > \pi/h$, then $\tilde{f}(w) = \hat{f}(w)$. However, if $f(x)$ is not band-limited, high-frequency components of $\hat{f}(w)$ are "folded" into the band $[-\pi/h, \pi/h]$ and they become indistinguishable from their low-frequency components or "alias." This phenomenon is known as aliasing.

The stability analysis of discrete-time evolving linear systems constitutes an important application of the concepts developed in this section.

Given a discrete function f_h, consider $S_+ f_h$ (the forward shifted sequence introduced in Definition A.15). Its discrete Fourier transform satisfies

$$\left(\widetilde{S_+ f_h}\right)(w) = \frac{h}{\sqrt{2\pi}} \sum_{j \in Z} f_{j+1} e^{-iwx_j} = \frac{h}{\sqrt{2\pi}} \sum_{j \in Z} f_j e^{-iwx_{j-1}}$$

$$= \frac{h}{\sqrt{2\pi}} \sum_{j \in Z} f_j e^{-iw(x_j - h)} = e^{iwh} \frac{h}{\sqrt{2\pi}} \sum_{j \in Z} f_j e^{-iwx_j} = e^{iwh} \tilde{f}_h(w).$$

This shows that in the "discrete" transform space, e^{iwh} is an "eigenfunction" for the operator S_+.

If we have a finite-difference operator A, linear with constant coefficients, A can always be expressed as

$$A = \left\{ \sum_{k=n_1}^{n_2} a_k (S_+)^k \right\}.$$

Now consider the relation

$$f_h^n = A f_h^{n-1}, \tag{A.7}$$

which represents the evolution of the solution of a linear system of partial differential equations from time $(n-1)\Delta t$ to time $n\Delta t = T$, with "initial data" f_h^0. Taking discrete Fourier transforms in (A.7), the evolution equation becomes

$$\left(\widetilde{f_h^n}\right)(w) = \rho(wh)\left(\widetilde{f_h^{n-1}}\right)(w). \tag{A.8}$$

The function $\rho(wh)$ is called the "symbol" of the operator A and it follows that if $\rho(wh)$ satisfies the Von Neumann condition

$$|\rho(wh)| \le 1 + c\,\Delta t, \tag{A.9}$$

with c a nonnegative real constant independent of w, h, and Δt, then by

Theorem A.15,

$$\|f_h^n\|_2^2 = \sum_k (f_k^n)^2 h = \int_{-\pi/h}^{\pi/h} \left|\left(\widetilde{f_h^n}\right)(w)\right|^2 dw.$$

Using (A.8),

$$\|f_h^n\|_2^2 = \int_{-\pi/h}^{\pi/h} [\rho(wh)]^2 \left|\left(\widetilde{f_h^{n-1}}\right)(w)\right|^2 dw,$$

and by Von Neumann's condition (A.9),

$$\|f_h^n\|_2^2 \le \int_{-\pi/h}^{\pi/h} [1 + c\,\Delta t]^2 \left|\left(\widetilde{f_h^{n-1}}\right)(w)\right|^2 dw$$

$$= [1 + c\,\Delta t]^2 \left\|\left(\widetilde{f_h^{n-1}}\right)\right\|^2$$

$$= [1 + c\,\Delta t]^2 \|f_h^{n-1}\|_2^2,$$

by Theorem A.15. Thus,

$$\|f_h^n\|_2 \le [1 + c\,\Delta t]\|f_h^{n-1}\|_2,$$

which implies

$$\|f_h^n\|_2 \le [1 + c\,\Delta t]^n \|f_h^0\|_2$$

$$\le e^{cn\,\Delta t}\|f_h^0\|_2 = e^{cT}\|f_h^0\|_2$$

and we have stability. ■

EXERCISE A.6

1. Show that the sequence of functions $f_j(x_p) = \cos jx_p$, $j = 0, 1, \ldots, m$, with the inner product $(f_j, f_n) = \sum_{p=1}^{m} f_j(x_p)f_n(x_p)$, $x_p = \pi(2p + 1)/(2(m + 1))$, is an orthogonal sequence.

2. Prove that for any two discrete functions $f_h = \{f_j\}_{j=0}^J$, $g_h = \{g_j\}_{j=0}^J$, $f_j = f(x_j)$, $g_j = g(x_j)$, $x_j = jh$, $Jh = L$,

$$(f_h, D_+g_h) = f_h g_h|_0^J - (g_h, D_-d_h).$$

3. Show that if $f_h = \{f_j\}_{j=0}^J$, $f_j = f(x_j)$, $x_j = jh$, $Jh = L$, is such that $f_j = 0$ for some j, $0 \le j \le J$, then

$$|f_j| \le L^{1/2}\|D_+f_h\|_2, \qquad j = 0, 1, \ldots, J \qquad \text{(Sobolev's inequality)}$$

and

$$\|f_h\|_2 \le L\|D_+f_h\|_2 \qquad \text{(Poincaré inequality)}.$$

4. Consider the heat equation $v_t = v_{xx}$, $t > 0$, $-\infty < x < \infty$, $v = \phi(x)$ at $t = 0$ and the finite-difference approximation with $u_h^n = \{u_j^n\}_{j=-\infty}^{\infty}$, $u_j^n = u(x_j, t_n) = u(jh, n\,\Delta t)$, given by

$$\frac{u_h^{n+1} - u_h^n}{\Delta t} = D_+D_-u_h^n.$$

Show that the method is stable if $\Delta t/h^2 \le 1/2$.

5. Do the stability analysis for the same problem as in Exercise 4 but for the Crank–Nicolson (trapezoidal rule) scheme:

$$u_h^{n+1} = u_h^n + \frac{\Delta t}{2}\left[D_+D_-u_h^n + D_+D_-u_h^{n+1}\right].$$

A.7   REFERENCES AND COMMENTS

The following references and comments serve to expand the basic material covered in the corresponding sections.

A.1–A.3. A systematic treatment of L^p spaces is available in

H. L. Royden, *Real Analysis*, Collin-Macmillan, London, 1968.

An introduction to L^p and Hilbert spaces at an elementary but rigorous level can be found in

K. Rektorys, *Variational Methods in Mathematics, Science and Engineering*, D. Reidel, Boston, 1980.

An excellent treatise of integration theory along classical lines is described in

V. I. Smirnov, *A Course of Higher Mathematics, Integration and Functional Analysis* **5**, Pergamon, Oxford, 1964.

For a more general overview of functional analysis and topology, see

E. Hille, *Methods in Classical and Functional Analysis*, Addison-Wesley, Reading, MA, 1972.

S. Lang, *Real Analysis*, Addison-Wesley, Reading, MA, 1969.

For more advanced study of measure theory and Lebesgue integration, see

P. R. Halmos, *Measure Theory*, Van Nostrand Reinhold, New York, 1950.

A.4. This section follows closely the description in Chap. 1 of

S. Agmon, *Lectures on Elliptic Boundary Value Problems*, Van Nostrand, Princeton, NJ, 1985.

See also Chap. 2 in

J. T. Oden and J. N. Reddy, *An Introduction to the Mathematical Theory of Finite Elements*, Wiley-Interscience, New York, 1976.

A.5. The practical character of Fourier transforms has been emphasized in the work of

A. Papoulis, *The Fourier Integral and Its Applications*, McGraw-Hill, New York, 1962.

The introduction of Whittaker's cardinal functions follows

R. Vichnevetsky and J. B. Bowles, *Fourier Analysis of Numerical Approximations of Hyperbolic Equations*, SIAM, Philadelphia, 1982.

A.6. For a good introduction to the subtleties and difficulties of the stability analysis of finite-difference schemes, see

R. Richtmyer and K. Morton, *Difference Methods for Initial-Value Problems*, Interscience, New York, 1967.

A more recent reference is the very readable textbook by

J. C. Strikwerda, *Finite Difference Schemes and Partial Differential Equations*, Wadsworth & Brooks/Cole, Pacific Grove, CA, 1988.

APPENDIX B

REFERENCES TO THE LITERATURE ON THE IHCP

Abdullaeva, G. Z. A certain inverse problem for the heat equation. *Akad. Nauk. Azerbaidzhan. SSR Dokl.* **32**, No. 6 (1976), 8–10 (Russian, Azerbaijani, and English summaries).

Abdullaeva, G. Z. Classical solvability in the large, of a nonlinear one-dimensional inverse boundary-value problem for second-order parabolic equations. *Akad. Nauk. Azerbaidzhan. SSR Dokl.* **37**, No. 11 (1981), 3–6 (Russian).

Abdullaeva, G. Z. Investigation of the classical solution of a nonlinear one-dimensional inverse boundary-value problem for second-order parabolic equations. *Akad. Nauk. Azerbaidzhan. SSR Dokl.* **38**, No. 3 (1982), 3–5 (Russian, Azerbaijani, and English summaries).

Ahundov, A. J. A multidimensional inverse problem for an equation of parabolic type in an unbounded domain. *Izv. Akad. Nauk. Azerbaidzhan. SSR Ser. Fiz.-Tehn. Mat. Nauk.*, No. 1 (1977), 86–91 (Russian, Azerbaijani, and English summaries).

Aleksakhin, A. A., and Ena, S. V. Effect of temperature measurement errors on the accuracy of boundary conditions. *J. Engrg. Phys.* **56**, No. 3 (1989), 275–277.

Aleksashenko, A. A. Analytic methods of solving inverse problems of heat and mass transfer. *Theoretical Foundation of Chemical Engineering* **18**, No. 2 (1984), 115–122.

Aleksashenko, A. A. Applicability of prediction formulae to inverse problems for heat and mass transfer. *Theoretical Foundation of Chemical Engineering* **23**, No. 3 (1990), 181–188.

Aliev, G. G. A certain inverse problem for the heat equation on a semi-infinite domain. *Izv. Akad. Nauk. Azerbaidzhan. SSR Ser. Fiz.-Tehn. Mat. Nauk.*, No. 7 (1970), 42–48 (Russian and Azerbaijani summaries).

Aliev, G. G. The problem of finding the functions that occur in the boundary and supplementary conditions in the inverse problem for the heat equation on a semi-infinite region. *Azerbaidzhan. Gos. Univ. Uchen. Zap. Ser. Fiz.-Mat. Nauk.*, No. 4 (1972), 25–31 (Russian and Azerbaijani summaries).

Aliev, G. G. A certain nonlinear inverse problem of nonstationary heat conduction. *Izv. Akad. Nauk. Azerbaidzhan. SSR Ser. Fiz.-Tehn. Mat. Nauk.*, No. 2 (1973), 121–128 (Russian and Azerbaijani summaries).

Alifanov, O. M. Inverse boundary value problems of heat conduction. *J. Engrg. Phys.* **29** (1975), 821–830.

Alifanov, O. M. Methods of solving ill-posed problems. *J. Engrg. Phys.* **45**, No. 5 (1984), 1237–1245.

Alifanov, O. M. *Inverse Heat Transfer Problems*. Springer-Verlag, to appear.

Alifanov, O. M., and Artyukhin, E. A. Regularized numerical solution of nonlinear inverse heat-conduction problem. *J. Engrg. Phys.* **29** (1975), 934–938.

Alifanov, O. M., and Egorov, Y. V. Algorithms and results of solving the inverse heat conduction boundary problem in a two-dimensional formulation. *J. Engrg. Phys.* **48**, No. 4 (1985), 489–496.

Alifanov, O. M., and Rumyantsev, S. V. Formulas for the discrepancy gradient in the iterative solution of inverse heat conduction problems. II. Determining the gradient in terms of a conjugate variable. *J. Engrg. Phys.* **52**, No. 4 (1987), 489–495.

Alifanov, O. M., and Rumyantsev, S. V. Application of iterative regularization for the solution of incorrect inverse problems. *J. Engrg. Phys.* **53**, No. 5 (1988), 1335–1342.

Alifanov, O. M., Artyukhin, E. A., and Rumyantsev, S. V. *Extremal Methods for Solving Incorrect Problems and Their Applications to Inverse Heat Mass Transfer Problems*. Nauka, Moscow, 1988 (Russian).

Al-Najem, N. M., and Ozisik, N. N. On the solution of three-dimensional inverse heat conduction problems in finite media. *Internat. J. Heat Mass Transfer* **28** (1985), 2121–2128.

Al-Najem, N. M., and Ozisik, M. N. Direct analytical approach for solving linear inverse heat conduction problems. *J. Heat Transfer, Trans. ASME* **107**, No. 3 (1985), 700–702.

Amonov, B. K., and Shishatskii, S. P. An a priori estimate of the solution of the Cauchy problem with data on a time-like surface for a second-order parabolic equation and related uniqueness theorems. *Dokl. Akad. Nauk. SSSR* **206** (1972), 11–12. English translation in *Soviet Math. Dokl.* **13**, No. 5 (1972), 1153–1154.

Anger, G., and Czerner, R. Solution of an inverse problem for the heat equation by methods of modern potential theory. In *Inverse and Improperly Posed Problems in Differential Equations (Proc. Conf., Math. Numer. Methods, Halle, 1979)*. Akademie-Verlag, Berlin, (1979) pp. 9–23; Math. Res. **1**.

Aripov, M. M., and Khaidarov, M. A Cauchy problem for a nonlinear heat equation in an inhomogeneous medium. *Dokl. Akad. Nauk. SSSR*, No. 2 (1986), 11–13 (Russian).

Arledge, R. G., and Haji-Sheikh, A. An iterative approach to the solution of inverse heat conduction problems. *Numer. Heat Transfer* **1** (1978), 365–376.

Artyukhin, E. A., and Rumyantsev, S. V. Optimal choice of descent steps in gradient methods of solution of inverse heat conduction problems. *J. Engrg. Phys.* **39**, No. 2 (1981), 865–869.

Artyukhin, E. A., Killikh, V. E., Nenarokomov, A. V., and Repin, I. V. Investigation of the thermal interaction of material with two-phase flows by the inverse problem method. *High Temperature* **28**, No. 1 (1990), 94–99.

Balakovskii, S. L. Resolving power of the iteration method of solving inverse heat conduction boundary value problems. *J. Engrg. Phys.* **52**, No. 6 (1987), 713–717.

Balakovskii, S. L. Solution of inverse heat transfer problems by a two-model method. *J. Engrg. Phys.* **57**, No. 3 (1990), 1118–1122.

Balk, P. I. Practically achievable accuracy and reliability of the solution of inverse heat conduction problems. *J. Engrg. Phys.* **52**, No. 2 (1987), 243–249.

Bass, B. R. Applications of the finite element to the inverse heat conduction problem using Beck's second method. *J. Engrg. Ind.* **102** (1980), 168–176.

Bass, B. R., and Ott, L. J. A finite element formulation of the two-dimensional inverse heat conduction problem. *Adv. in Comput. Tech.* **2** (1980), 238–248.

Baumeister, J. On the treatment of free-boundary problems with the heat equation via optimal control. *Math. Methods Appl. Sci.* **1**, No. 1 (1979), 1–61.

Baumeister, J. *Stable Solution of Inverse Problems.* Vieweg and Sohn, Braunschweig/Wiesbaden, 1987.

Baumeister, J., and Reinhardt, H. J. Inverse heat conduction problems and their numerical solution. In *Inverse and Ill-Posed Problems*, H. W. Engl and C. W. Groetsch (eds.). Academic, Boston, 1987, pp. 325–344.

Beck, J. V. Correction of transient thermocouple temperature measurements in heat-conducting solids. II. The calculation of transient heat fluxes using the inverse convolution. Technical Report RADTR-7-60-38 (Part II), Res. and Adv. Dev. Div., AVCO Corp., Wilmington, MA, March 30, 1961.

Beck, J. V. Calculation of surface heat flux from an internal temperature history. ASME Paper 62-HT-46, 1962.

Beck, J. V. Surface heat flux determination using an integral method. *Nucl. Engrg. Des.* **7** (1968), 170–178.

Beck, J. V. Nonlinear estimation applied to the nonlinear heat conduction problem, *Internat. J. Heat Mass Transfer* **13** (1970), 703–716.

Beck, J. V. Criteria for comparison of methods of solution of the inverse heat conduction problem. *Nucl. Engrg. Des.* **53** (1979), 11–22.

Beck, J. V., and Murio, D. A. Combined function specification–regularization procedure for solution of inverse heat conduction problem. *AIAA J.* **24**, No. 1 (1986), 180–185.

Beck, J. V., and Wolf, H. The nonlinear inverse heat conduction problem. ASME Paper 65-HT-40, presented at the ASME/AIChE Heat Transfer Conference and Exhibit, Los Angeles, August 8–11, 1965.

Beck, J. V., Blackwell, B., and St. Clair, Ch. R., Jr. *Inverse Heat Conduction. Ill-Posed Problems.* Wiley-Interscience, New York, 1985.

Beck, J. V., Litkouhi, B., and St. Clair, C. R., Jr. Efficient sequential solution of the nonlinear inverse heat conduction problem. *Numer. Heat Transfer* **5** (1982), 275–286.

Bell, J. B. The noncharacteristic Cauchy problem for a class of equations with time dependence. I. Problems in one space dimension. *SIAM J. Math. Anal.* **12**, No. 5 (1981), 759–777.

Bell, J. B. The noncharacteristic Cauchy problem for a class of equations with time dependence. II. Multidimensional problems. *SIAM J. Math. Anal.* **12**, No. 5 (1981), 778–797.

Bell, J. B., and Wardlaw, A. B. Numerical solution of an ill-posed problem arising in wind tunnel heat transfer data reduction. Naval Surface Weapons Center, NSWC TR 82-32, December 1981.

Ben-Haim, Y., and Elias, E. Indirect measurements of surface temperature and heat flux: optimal design using convexity analysis. *Internat. J. Heat Mass Transfer* **30**, No. 8 (1987), 1673–1683.

Beznoshchenko, N. J., and Prilepko, A. I. Inverse problems for equations of parabolic type. In *Problems in Mathematical Physics and Numerical Mathematics*. Nauka, Moscow, 1977, pp. 51–63 (Russian).

Blackwell, B. F. An efficient technique for the numerical solution of the one-dimensional inverse problem of heat conduction. *Numer. Heat Transfer* **4** (1981), 229–239.

Blackwell, B. F. Some comments on Beck's solution of the inverse problem of heat conduction through the use of Duhamel's theorem. *Internat. J. Heat Mass Transfer* **26** (1983), 302–305.

Burggraf, O. R. An exact solution of the inverse problem in heat conduction theory and applications. *J. Heat Transfer* **86C** (1964), 373–382.

Busby, H. R., and Trujillo, D. M. Numerical solution to a two-dimensional inverse heat conduction problem. *Internat. J. Numer. Methods in Engrg.* **21** (1985), 349–359.

Cannon, J. R. A Cauchy problem for the heat equation. *Ann. Mat. Pura Appl.* (4) **66** (1964), 155–165.

Cannon, J. R. A priori estimate for continuation of the solution of the heat equation in the space variable. *Ann. Mat. Pura Appl.* (4) **65** (1964), 377–387.

Cannon, J. R. Determination of an unknown heat source from overspecified boundary data. *SIAM J. Numer. Anal.* **5** (1968), 275–286.

Cannon, J. R. *The One-Dimensional Heat Equation*. Addison-Wesley, Menlo Park, CA, 1984.

Cannon, J. R., and Douglas, J., Jr. The Cauchy problem for the heat equation. *SIAM J. Numer. Anal.* **4** (1967), 317–336.

Cannon, J. R., and Douglas, J., Jr. The approximation of harmonic and parabolic functions on half-spaces from interior data. In *Numerical Analysis of Partial Differential Equations* (C.I.M.E. 2 Ciclo, Ispra, 1967). Edizioni Cremonese, Rome, 1968, pp. 193–230.

Cannon, J. R., and DuChateau, P. Determining unknown coefficients in a nonlinear heat-conduction problem. *SIAM J. Appl. Math.* **24** (1973), 298–314.

Cannon, J. R., and DuChateau, P. An inverse problem for an unknown source in a heat equation. *J. Math. Anal. Appl.* **75**, No. 2 (1980), 465–485.

Cannon, J. R., and Ewing, R. E. A direct numerical procedure for the Cauchy problem for the heat equation. *J. Math. Anal. Appl.* **56** (1976), 7–17.

Cannon, J. R., and Hill, C. D. Continuous dependence of bounded solutions of a linear, parabolic partial-differential equation upon interior Cauchy data. *Duke Math. J.* **35** (1968), 217–230.

Cannon, J. R., and Knightly, G. H. The approximation of the solution of the heat equation in a half-strip from data specified on the bounding characteristics. *SIAM J. Numer. Anal.* **6** (1969), 149–159.

Cannon, J. R., and Zachmann, D. Parameter determination in parabolic partial-differential equations from overspecified boundary data. *Internat. J. Engrg. Sci.* **20**, No. 6 (1982), 779–788.

Carasso, A., Determining surface temperatures from interior observations. *SIAM J. Appl. Math.* **42**, No. 3 (1982), 558–574.

Carasso, A. Nonlinear inverse heat transfer calculations in gun barrels, ARO Report 84-1, (1984).

Carasso, A. Infinitely divisible pulses, continuous deconvolution, and the characterization of linear time invariant systems. *SIAM J. Appl. Math.* **47** (1987), 892–927.

Carasso, A. S. Impulse response acquisition as an inverse heat conduction problem. *SIAM J. Appl. Math.* **50**, No. 1 (1990), 75–90.

Carasso, A. S. Corrigendum: Impulse response acquisition as an inverse heat conduction problem. *SIAM J. Appl. Math.* **50**, No. 3 (1990), 942.

Carasso, A. Space marching difference schemes in the nonlinear inverse heat conduction problem. Preprint NISTIR 4482, U.S. Department of Commerce, 1990.

Carasso, A., and Hsu, N. N. Probe waveforms and deconvolution in the experimental determination of elastic Green's functions. *SIAM J. Appl. Math.* **45** (1985), 396–482.

Carasso, A., and Hsu, N. N. L^{∞} error bounds in partial deconvolution of the inverse Gaussian pulse. *SIAM J. Appl. Math.* **45** (1985), 1029–1038.

Carasso, A., Sanderson, J. G., and Hyman, J. M. Digital removal of random media image degradations by solving the diffusion equation backwards in time. *SIAM J. Numer. Anal.* **15** (1978), 344–367.

Casas, E. The Cauchy problem for the heat equation in a half space. Mathematical contributions in memory of Professor Víctor Manuel Onieva Aleixandre, University of Cantabria, Santander, 1991, pp. 95–103.

Chen, H., and Chang, S. Application of the hybrid method to inverse heat conduction problems. *Internat. J. Heat Mass Transfer* **33**, No. 4 (1990), 621–628.

Chewning, W. C., and Seidman, T. I. A convergent scheme for boundary control of the heat equation. *SIAM J. Control Optim.* **15**, No. 1 (1977), 64–72.

Chrzanowski, E. M. An inverse problem for a combined system of diffusion equations. *Demonstratio Math.* **14**, No. 2 (1981), 427–436.

Colton, D. The noncharacteristic Cauchy problem for parabolic equations in two space variables. *Proc. Amer. Math. Soc.* **41** (1973), 551–556.

Colton, D. The inverse Stefan problem. *Ber. Gesell. Math. Datenverarbeitung Bonn*, No. 77 (1973), 29–41.

Colton, D. The inverse Stefan problem for the heat equation in two space variables. *Mathematika* **21** (1974), 282–286.

Colton, D. The noncharacteristic Cauchy problem for parabolic equations in one space variable. *SIAM J. Math. Anal.* **5** (1974), 263–272.

Colton, D. Continuation and reflection of solutions to parabolic partial-differential equations. In *Ordinary and Partial-Differential Equations (Proc. Fifth Conf., Univ. Dundee, 1978). Lecture Notes in Math.* **827**. Springer, Berlin, 1980, pp. 54–82.

Cozdoba, L. A., and Crykowsky, P. G. *Methods of Solution to the Inverse Problem.* Scientific Publisher, Kiev, 1982 (Russian).

Davies, A. R. On a constrained Fourier extrapolation method for numerical deconvolution. In *Improperly Posed Problems and Their Numerical Treatment*, G. Hammerlin and K.-H. Hoffman (eds.). Birkhauser-Verlag, Basel, 1983, pp. 65–80.

Davies, J. M. Input power determined from temperatures in simulated skin protected against thermal radiation. *J. Heat Transfer* **99** (1966), 154–160.

Denisov, A. M. On the approximate solution of a Volterra equation of the first kind, related to an inverse problem for the heat equation. *Vestnik Moskov. Univ. Ser. XV Vychisl. Mat. Kibernet.*, No. 3 (1980), 49–52 (Russian).

Denisov, A. M. Uniqueness of the solutions of some inverse problems for the heat equation with a piecewise-constant coefficient. *Zh. Vychisl. Mat. i Mat. Fiz.* **22**, No. 4 (1982), 858–864 (Russian).

Denisov, V. N. Stabilization of the solution of the Cauchy problem for the heat equation. *Soviet Math. Dokl.* **37**, No. 3 (1988), 688–692.

Dikinov, H. Z., Kerefov, A. A., and Nahushev, A. M. A certain boundary-value problem for a loaded heat equation. *Differencial'nye Uravnenija* **12**, No. 1 (1976), 191–292 (Russian).

Douglas, J., Jr. Approximate continuation of harmonic and parabolic functions. In *Numerical Solution of Partial Differential Equations (Proc. Symp. Univ. Maryland, 1965)*. Academic, New York, 1966, pp. 353–364.

Drentchev, L. B. An analytic solution for one inverse problem of heat transfer. *C. R. Acad. Bulgare Sci.* **42** (1989), 35–38.

D'Souza, N. Inverse heat conduction problem for prediction of surface temperatures and heat transfer from interior temperature measurements. Report SRC-R-74, Space Research Corporation, Montreal, December 1973.

D'Souza, N. Numerical solution of one-dimensional inverse transient heat conduction by finite difference method. ASME Paper 75-WA/HT-81, presented at Winter Annual Meeting, Houston, November 30–December 4, 1975.

Duc, V. T., Hao, D. N., Ngoc, N. T., and Gorenflo, R. On the Cauchy problems for systems of partial differential equations with a distinguished variable. *Numer. Funct. Anal. Optim.* **12** (1991), 213–236.

Edelstein, W. S. Further study of spatial-decay estimates for semilinear parabolic equations. *J. Math. Anal. Appl.* **35** (1971), 577–590.

Effros, E. G., and Kazdan, J. L. On the Dirichlet problem for the heat equation. *Indiana Univ. Math. J.* **20** (1970/1971), 683–693.

Egorov, A. I. Conditions for optimality in a certain problem of control of a heat-transfer process. *Zh. Vychisl. Mat. i Mat. Fiz.* **12** (1972), 791–799 (Russian).

Egorov, A. I., and Naval, E. S. Optimal control of a process of heat conduction. *Prikl. Mat. i Programmirovanie* **13** (1975), 44–63 (Russian).

Eldén, L. The numerical solution of a non-characteristic Cauchy problem for a parabolic equation. In *Numerical Treatment of Inverse Problems in Differential and Integral Equations*, P. Deuflhard and E. Hainer (eds.). Birkhauser, Boston, 1983, pp. 246–268.

Eldén, L. Approximations for a Cauchy problem for the heat equation. *Inverse Problems* **3** (1987), 263–273.

Eldén, L. Modified equations for approximating the solution of a Cauchy problem for the heat equation. In *Inverse and Ill-Posed Problems*, H. W. Engl and C. W. Groetsch (eds.). Academic, Boston, 1987, pp. 345–350.

Eldén, L. Hyperbolic approximations for a Cauchy problem for the heat equation. *Inverse Problems* **4** (1988), 59–70.

Engl, H., and Langthaler, T. Numerical solution of an inverse problem connected with continuous casting of steel. *ZOR* **29** (1985), 185–199.

Engl, H., and Manselli, P. Stability estimates and regularization for an inverse heat conduction problem. *Numer. Funct. Anal. Optim.* **10** (1989), 517–540.

Engl, H., Langthaler, T., and Manselli, P. On an inverse problem for a nonlinear heat equation connected with continuous casting of steel. In *Optimal Control with Partial Differential Equations II*, K.-H. Hoffmann and W. Krabs (eds.). Birkhauser, Basel, 1987, pp. 67–89.

Ewing, R. E. The Cauchy problem for a linear parabolic partial differential equation. *J. Math. Anal. Appl.* **71** (1979), 167–186.

Ewing, R. E., and Falk, R. S. Numerical approximation of a Cauchy problem for a parabolic partial differential equation. *Math. Comp.* **33** (1979), 1125–1144.

Fasano, A. Un esempio controllo ottimale in un problema del tipo di Stefan, *Boll. Un. Mat. Ital.* (4) **4** (1971), 846–858 (Italian and English summaries).

Fattorini, H. O. The time-optimal problem for boundary control of the heat equation. In *Calculus of Variations and Control Theory (Proc. Symp., Math. Res. Center, Univ. Wisconsin, Madison, WI, 1975*; dedicated to Laurence Chosholm Young on the occasion of his 70th birthday). Academic, New York, 1976, pp. 305–320.

Fattorini, H. O. Reachable states in boundary control of the heat equation are independent of time. *Proc. Roy. Soc. Edinburgh Sect. A* **81**, Nos. 1–2 (1978), 71–77.

Flach, G. P., and Ozisik, M. N. Whole domain function-specification method for linear inverse heat conduction. *Ann. Nucl. Energy* **13**, No. 6 (1986), 325–336.

Flach, G. P., and Ozisik, M. N. Inverse heat conduction of periodically contacting surfaces. *J. Heat Transfer, Trans. ASME* **110**, No. 4 (1988), 821–829.

Flach, G. P., and Ozisik, M. N. Inverse heat conduction problem of simultaneously estimating spatially varying thermal conductivity and heat capacity per unit volume. *Numer. Heat Transfer* **16**, No. 2 (1989), 249–266.

France, D. M., and Chiang, T. Analytical solution to inverse heat conduction problems with periodicity. *J. Heat Transfer* **102** (1980), 579–581.

Frank, I. An application of least square methods to the solution of the inverse problem of heat conduction. *J. Heat Transfer* **85** (1963), 378–379.

Garifo, I., Schrock, V. E., and Spedicato, E. On the solution of the inverse heat conduction problem by finite differences. *Energia Nucleare* **22** (1975), 452–464.

Gentili, G. Dissipativity condition and inverse problems for heat conduction with linear memory. *Inverse Problems* **7**, No. 1 (1991), 77–84.

Gillian, D. S., Mair, B. A., and Martin, C. F. A convolution method for inverse heat conduction problems. *Math. Systems Theory* **21**, No. 1 (1988), 49–60.

Gillian, D. S., Mair, B. A., and Martin, C. F. An inverse convolution method for regular parabolic equations. *SIAM J. Control Optim.* **29**, No. 1 (1991), 71–88.

Ginsberg, F. On the Cauchy problem for the one-dimensional heat equation. *Math. Comp.* **17** (1963), 257–269.

Giurgiu, M. A feedback solution of a linear quadratic problem for boundary control of heat equation. *Rev. Roumaine Math. Pures Appl.* **20**, No. 8 (1975), 927–954.

Giurgiu, M. Linear feedback for optimal stabilization by boundary control of heat equation. *Rev. Roumaine Math. Pures Appl.* **22**, No. 6 (1977), 777–796.

Glagoleva, R. J. The continuous dependence on the initial data of the solution of the first boundary value problem for a parabolic equation with negative time. *Dokl. Akad. Nauk. SSSR* **148** (1963), 20–23 (Russian).

Glasko, V. B., Kulik, N. N., Tikhonov, A. N., and Shkliano, I. N. An inverse problem of heat condition. *U.S.S.R. Computational Math. and Math. Phys.* **19** (1980), 223–230.

Glasko, V. B., Zakharov, M. V., and Kolp, A. Y. An application of the regularization method to solve an inverse problem of non-linear heat-conduction theory. *U.S.S.R. Computational Math. and Math. Phys.* (1975), 244–248.

Grysa, K. Methods of potential theory in solving inverse problems of heat conduction. *Mech. Teoret. Stos.* **20**, Nos. 3–4 (1983), 207–223 (Polish).

Grysa, K., and Cialkowsky, M. J. Inverse problems of temperature fields. A survey. *Mech. Teoret. Stos.* **18**, No. 4 (1980), 535–554.

Grysa, K., and Kaminski, H. On a time step choice in solving inverse heat conduction problems. *Z. Angew. Math. Mech.* **66**, No. 5 (1986), 368–370.

Grysa, K., Cialkowski, M. J., and Kaminski, H. An inverse temperature field problem of the theory of thermal stresses. *Nucl. Engrg. Des.* **64** (1981), 169–184.

Guo, L., and Murio, D. A mollified space marching finite difference algorithm for the two-dimensional inverse heat conduction problem with slab symmetry. *Inverse Problems* **7** (1991), 247–259.

Guo, L., Murio, D. A., and Roth, C. A mollified space marching finite differences algorithm for the inverse heat conduction problem with slab symmetry. *Comput. Math. Appl.* **19**, No. 7 (1990), 15–26.

Hagin, F. A stable approach to solving one-dimensional inverse problems. *SIAM J. Appl. Math.* **40** (1981), 439–453.

Hào, D. N. A non-characteristic Cauchy problem for linear parabolic equations. I. Solvability. Preprint A-91-36, Freie Universität Berlin, 1991.

Hào, D. N. A non-characteristic Cauchy problem for linear parabolic equations. II. A variational method. Preprint A-91-37, Freie Universität Berlin, 1991.

Hào, D. N. A non-characteristic Cauchy problem for linear parabolic equations. III. A variational method and its approximation schemes. Preprint A-91-38, Freie Universität Berlin, 1991.

Hào, D. N. A non-characteristic Cauchy problem for linear parabolic equations and related inverse problems. I. Solvability. Preprint A-91-39, Freie Universität Berlin, 1991.

Hào, D. N. A non-characteristic Cauchy problem for linear parabolic equations and related inverse problems. II. A variational method. Preprint A-91-40, Freie Universität Berlin, 1991.

Hào, D. N. Regularizing a noncharacteristic Cauchy problem for the heat equation. Preprint A-90-13, Freie Universität Berlin, 1991.

Hào, D. N. A mollification method for ill-posed problems. Preprint A-91-35, Freie Universität Berlin, 1991.

Hào, D. N., and Gorenflo, R. A non-characteristic Cauchy problem for the heat equation. *Acta Appl. Math.* **24** (1991), 1–27.

Hensel, E. *Inverse Theory and Applications in Engineering*. Prentice-Hall, Englewood Cliffs, NJ, 1990.

Hensel, E. C., and Hills, R. G. A space marching finite difference algorithm for the one dimensional inverse heat conduction transfer problem. ASME Paper 84-HT-48, presented at 22nd National Heat Transfer Conference, Niagara Falls, NY, August 6–8, 1984.

Hill, J., and Denson, C. Parabolic equations in one space variable and the noncharacteristic Cauchy problem. *Comm. Pure Appl. Math.* **20** (1967), 617–633.

Hills, R. G., and Hensel, E. C. One-dimensional nonlinear inverse heat conduction technique. *Numer. Heat Transfer* **10** (1986), 369–393.

Hills, R. G., and Mulholland, G. P. Accuracy and resolving power of one-dimensional transient inverse heat conduction theory as applied to discrete and inaccurate measurements. *Internat. J. Heat Mass Transfer* **22** (1979), 1221–1229.

Hills, R. G., Raynaud, M., and Hensel, E. Surface variance estimates using an adjoint formulation for one dimensional nonlinear inverse heat conduction technique. *Numer. Heat Transfer* **10** (1986), 441–461.

Howse, T. K. J., Kent, R., and Rawson, H. The determination of glass-mould heat fluxes from mould temperature measurements. *Glass Tech.* **12** (1971), 91–93.

Hsieh, C. K., and Su, K. C. A methodology of predicting cavity geometry based on the scanned surface temperature data—prescribed surface temperature at the cavity side. *J. Heat Transfer* **192** (1980), 324–329.

Hsieh, C. K., and Su, K. C. A methodology of predicting cavity geometry based on the scanned surface temperature data—prescribed heat flux at the cavity side. *J. Heat Transfer* **103** (1981), 42–46.

Hurd, A. E. Backward lower bounds for solutions of mixed parabolic problems. *Michigan Math. J.* **17** (1970), 97–102.

Imber, M. A temperature extrapolation method for hollow cylinders. *AIAA J.* **11** (1973) 117–118.

Imber, M. A temperature extrapolation mechanism for two-dimensional heat flow. *AIAA J.* **12** (1974), 1087–1093.

Imber, M. Two-dimensional inverse conduction problem—further observations. *AIAA J.* **13** (1975), 114–115.

Imber, M. Comments on: On transient cylindrical surface heat flux predicted from interior temperature responses. *AIAA J.* **14** (1975), 542–543.

Imber, M. Inverse problem for a solid cylinder. *AIAA J.* **17** (1979) 91–94.

Imber, M. Nonlinear heat transfer in planar solids: direct and inverse applications. *AIAA J.* **17** (1979), 204–212.

Imber, M., and Khan, J. Prediction of transient temperature distributions with embedded thermocouples. *AIAA J.* **10** (1972), 784–789.

Ishankulov, T. I., and Makhmudov, O. I. On the Cauchy problem for the heat equation. *Problems in Mathematical Analysis and Its Applications*. Samarkand Gos. Univ., Samarkand, 1986, pp. 46–50 (Russian).

Iskenderov, A. D. Inverse boundary-value problems with unknown coefficients for certain quasilinear equations. *Dokl. Akad. Nauk. SSSR* **178** (1968), 999–1002 (Russian).

Iskenderov, A. D. Certain multidimensional inverse boundary-value problems. *Azerbaidzhan. Gos. Univ. Uchen. Zap. Ser. Fiz.-Mat. Nauk.*, No. 2 (1968), 76–80 (Russian).

Ivanov, V. K. Ill-posed problems and deconverged processes. *Uspechi Mat. Nauk.* **11**, No. 4 (1985). 165–166. English translation in *Russian Math. Surveys* **40**, No. 4 (1985), 187–188.

Ivanov, V. K. Conditions for well-posedness in the Hadamard sense in spaces of generalized functions. *Sibrsk. Mat. Zh.* **28** (1987), 53–59. English translation in *Siberian Math. J.* **28** (1987), 906–911.

Jahno, V. G. An inverse problem for a system of parabolic equations. *Differencial'nye Uravnenija* **15**, No. 3 (1979), 566–569, 576 (Russian).

Jarny, Y., Ozisik, M. N., and Bardon, J. P. General optimization method using adjoint equation for solving multidimensional inverse heat conduction. *Internat. J. Heat Mass Transfer* **34**, No. 11 (1991), 2911–2919.

Jochum, P. The numerical solution of the inverse Stefan problem. *Numer. Math.* **34**, No. 4 (1980), 411–429.

Jochum, P. The inverse Stefan problem as a problem of nonlinear approximation theory. *J. Approx. Theory.* **30**, No. 2 (1980), 81–98.

Johnson, R. A priori estimates and unique continuation theorems for second-order parabolic equations. *Trans. Amer. Math. Soc.* **158** (1971), 167–177.

Kawohl, B. On a nonlinear heat-control problem with boundary conditions changing in time. *Z. Angew. Math. Mech.* **61**, No. 5 (1981), 248–249.

Keltner, N. R., and Beck, J. V. Surface temperature measurement errors. *J. Heat Transfer* **105** (1983), 312–318.

Klibanov, M. V. Some inverse problems for parabolic equations. *Mat. Zametki* **30**, No. 2 (1981), 203–210, 314 (Russian).

Knabner, P. Regularizing the Cauchy problem for the heat equation by sign restrictions. In *Improperly Posed Problems and Their Numerical Treatment*, G. Hammerlin and K.-H. Hoffmann (eds.). Birkhauser, Boston, 1983, pp. 165–177.

Knabner, P. Regularization of the Cauchy problem for the heat equation by norm bounds. *Appl. Anal.* **17** (1984) 295–312.

Knabner, P., and Vessella, S. Stability estimates for ill-posed Cauchy problems for parabolic equations. In *Inverse and Ill-Posed Problems*, H. W. Engl and C. W. Groetsch (eds.). Academic, Boston, 1987, pp. 351–368.

Knabner, P., and Vessella, S. Stabilization of ill-posed Cauchy problems for parabolic equations. *Ann. Mat. Pura Appl.* (4) **149** (1987), 393–409.

Knabner, P., and Vessella, S. The optimal stability estimate for some ill-posed Cauchy problems for a parabolic equation. *Math. Methods Appl. Sci.* **10** (1988), 575–583.

Kononova, A. A. The Cauchy problem for the inverse heat equation in an infinite strip. *Ural. Gos. Univ. Mat. Zap.* **8**, No. 4 (1974), pp. 58–63, 134 (Russian).

Kononova, A. A. On the problem of control for the heat equation in the uniform metric. *Ural. Gos. Univ. Mat. Zap.* **10**, No. 2, *Issled. Sovremen. Mat. Anal.* (1977), pp. 74–82, 216 (Russian).

Kover'yanov, V. A. Inverse problem of nonsteady-state thermal conductivity. *Teplofizika Vysokikh Temperatur* **5**, No. 1 (1967), 141–143.

Kuroyanagi, T. Surface temperature and surface heat flux determination of the inverse heat conduction problem. *Bull. JSME* **29**, No. 255 (1986), 2961–2969.

Kuroyanagi, T. Transformation of Duhamel's integral in the inverse heat conduction problem. *Bull. JSME* **29**, No. 255 (1986), 2953–2960.

Kurpisz, K. Method for determining steady state temperature distribution within blast furnace hearth lining by measuring temperature at selected points. *Trans. Iron and Steel Institute of Japan*, **28**, No. 11 (1988), 926–929 (Japanese).

Kurpisz, K. Numerical solution of one case of inverse heat conduction problems. *J. Heat Transfer, Trans. ASME* **113**, No. 2 (1991), 280–286.

Lanconelli, E. Sul problema de Dirichlet per l'equazione del calore. *Ann. Mat. Pura Appl.* (4) **97** (1973), 83–114 (Italian).

Langford, D. New analytic solutions of the one-dimensional heat equation for temperature and heat flow rate both prescribed at the same fixed boundary (with applications to the phase change problem). *Quart. Appl. Math.* **24** (1976), 315–322.

Lavrent'ev, M. M., and Amonov, B. K. Determination of the solutions of the diffusion equation from its values on discrete sets. *Dokl. Akad. Nauk. SSSR* **228**, No. 6 (1976), 1284–1285 (Russian).

Lavrent'ev, M. M., and Klibanov, M. V. A certain integral equation of the first kind, and the inverse problem for a parabolic equation. *Dokl. Akad. Nauk. SSSR* **221**, No. 4 (1975), 782–783 (Russian).

Lavrent'ev, M. M., and Klibanov, M. V. A certain inverse problem for an equation of parabolic type. *Differencial'nye Uravnenija* **11**, No. 9 (1975), 1647–1651, 1717 (Russian).

Lavrent'ev, M. M. Romanov, V. G., and Shisatskii, S. P. *Ill-Posed Problems of Mathematical Physics and Analysis. Trans. Math. Monographs* **64**. AMS, Providence, 1986.

Lazuchenkov, N. M., and Shmukia, A. A. Solution of a two-dimensional inverse problem for a differential equation of parabolic type. In *Differential Equations and Their Applications*. Dnepropetrovsk Gos. Univ., Dnepropetrovsk, 1976, pp. 96–100 (Russian).

Lees, M., and Protter, M. H. Unique continuation for parabolic differential equations and inequalities. *Duke Math. J.* **28** (1961), 369–382.

Levine, H. A. Continuous data dependence, regularization and a three lines theorem for the heat equation with data in a space like direction. *Ann. Math. Pura Appl.* (4) **134** (1983), 267–286.

Levine, H. A., and Vessella, S. Estimates and regularization for solutions of some ill-posed problems of elliptic and parabolic type. *Rend. Circ. Mat. Palermo* **123** (1980), 161–183.

Lin, D. Y. T., and Westwater, J. W. Effect of metal thermal properties on boiling curves obtained by the quenching method. *Heat Transfer 1982—Munchen Conf. Pro.* **4**. Hemisphere, New York, 1982, pp. 155–160.

Lions, J. L. Sur la stabilisation de certains problems mal pośes. *Rend. Sem. Mat. Fis. Milano* **36** (1966), 80–87 (English summary).

Liu, G. L. A novel variational formulation of inverse problem of heat conduction with free boundary on an image plane. *Numerical Methods in Thermal Problems* **6**, Parts 1 and 2. Pineridge, Swansea, 1989, pp. 1712–1720.

Louis, A. K., and Maass, P. A mollifier method for linear operator equations of the first kind. *Inverse Problems* **6** (1990), 427–490.

Louis, A. K., and Maass, P. Smoothed projection methods for the moment problem. *Numer. Math.* **59** (1991), 277–294.

Maillet, D., Degiovanni, A., and Pasquetti, R. Inverse heat conduction applied to the measurement of heat transfer coefficient on a cylinder. Comparison between an analytical and a boundary element technique. *J. Heat Transfer, Trans. ASME* **113**, No. 3 (1991), 549–557.

Mair, B. A. On the recovery of surface temperature and heat flux via convolution. *Computation and Control, Bozeman, MT, 1988.* Birkhauser, Boston, 1989, pp. 197–207.

Malishev, I. G. A certain inverse problem for a parabolic equation with a variable coefficient. *Visnik Kiiv. Univ. Ser. Mat. Meh.*, No. 16 (1974), 32–37, 173 (Ukrainian, English, and Russian summaries).

Malishev, I. G. Reduction of the inverse problem for a parabolic equation to the spectral-inverse problem of the Schrodinger equation. *Visnik Kiiv. Univ. Ser. Mat. Meh.*, No. 18 (1976), 125–129, 145 (Ukrainian, English, and Russian summaries).

Malyshev, I. G. The second inverse problem for the heat equation in a non cylindrical domain. *Vychisl. Prikl. Mat. (Kiev)*, No. 25 (1975), 56–64, 144 (Russian and English summaries).

Malyshev, I. G. Inverse problems for the heat equation in a domain with a moving boundary. *Ukrain. Mat. Zh.* **27**, No. 5 (1975), 56–64, 144 (Russian).

Malyshev, I. G., and Malysheva, G. M. The approximate determination of the coefficients in the heat equation. *Vychisl. Prikl. Mat. (Kiev)* **32** (1977), 132–136, 159 (Russian and English summaries).

Manselli, P., and Miller, K. Calculation of the surface temperature and heat flux on one side of a wall from measurements on the opposite side. *Ann. Mat. Pura Appl.* (4) **123** (1980), 161–183 (English and Italian summaries).

Markin, A. D., and Pyatyshkin, G. G. Regularization of the solution of the inverse heat conduction problem in a variational formulation. *J. Engrg. Phys.* **52**, No. 6 (1987), 717–721.

Marquardt, W., and Auracher, H. Observer-based solution of inverse heat conduction problems. *Internat. J. Heat Mass Transfer* **33**, No. 7 (1990), 1545–1562.

Maslov, V. P. Regularization of incorrect problems for singular integral equations. *Dokl. Akad. Nauk. SSSR* **176**, No. 5 (1967). English translation in *Soviet Math. Dokl.* **8**, No. 5 (1967), 1251–1254.

Masuda, K. A note on the analyticity in time and the unique continuation property for solutions of diffusion equations. *Proc. Japan Acad.* **43** (1967), 420–422.

Mehta, R. C. Extension of the solution of inverse conduction problem. *Internat. J. Heat Mass Transfer* **22** (1979), 1149–1150.

Miller, K. Three circle theorems in partial-differential equations and applications to improperly posed problems. *Arch. Rational Mech. Anal.* **16** (1964), 126–154.

Miller, K., and Viano, G. On the necessity of nearly-best possible methods for analytic continuation of scattering data. *J. Math. Phys.* **14**, No. 8 (1973), 1037–1048.

Miranker, W. E. A well-posed problem for the backward heat equation. *Proc. Amer. Math. Soc.* **12** (1961), 243–247.

Mizel, V. J., and Seidman, T. I. Observation and prediction for the heat equation. II. *J. Math. Anal. Appl.* **38** (1972), 149–166.

Mizohata, S. Unicité du prolongement des solutions pour quelques operateurs differentiels paraboliques. *Mem. Coll. Sci. Univ. Kyoto. Ser. A Math.* **31** (1958), 219–239.

Monk, P. Error estimates for a numerical method for an ill-posed Cauchy problem for the heat equation. *SIAM J. Numer. Anal.* **23** (1986), 1155–1172.

Mulholland, G. P., and San Martin, R. L. Indirect thermal sensing in composite media. *Internat. J. Heat Mass Transfer* **16** (1973), 1056–1060.

Mulholland, G. P., Gupta, B. P., and San Martin, R. L. Inverse problem of heat conduction in composite media. ASME Paper 75-WA/HT-83, 1975.

Murio, D. A. Numerical methods for inverse, transient heat-conduction problems. *Rev. Un. Mat. Argentina* **30**, No. 1 (1981), 25–46.

Murio, D. A. The mollification method and the numerical solution of an inverse heat conduction problem. *SIAM J. Sci. Statist. Comput.* **2** (1981), 17–34.

Murio, D. A. On the estimation of the boundary temperature on a sphere from measurements at its center. *J. Comput. Appl. Math.* **8**, No. 2 (1982), 111–119.

Murio, D. A. On the characterization of the solution of the inverse heat conduction problem. Paper 85-WA/HT-41, presented at the 1985 ASME Meeting, Session on Inverse and Optimization Heat Transfer, Miami Beach, FL.

Murio, D. A. Parameter selection by discrete mollification and the numerical solution of the inverse heat conduction problem. *J. Comput. Appl. Math.* **22** (1988), 25–34.

Murio, D. A. On the noise reconstruction for the inverse heat conduction problem. *Comput. Math. Appl.* **16**, No. 12 (1988), 1027–1033.

Murio, D. A. The mollification method and the numerical solution of the inverse heat conduction problem by finite differences. *Comput. Math. Appl.* **17**, No. 10 (1989), 1385–1396.

Murio, D. A. Numerical identification of boundary conditions on nonlinearly radiating inverse heat conduction problems. In *Inverse Design Concepts and Optimization in Engineering Sciences*, G. Dulikravich (ed.). Washington DC, 1991, pp. 227–238.

Murio, D. A. Solution of inverse heat conduction problems with phase changes by the mollification method. *Comput. Math. Appl.* **24**, No. 4 (1992), 45–57.

Murio, D. A., and Guo, L. A stable marching finite difference algorithm for the inverse heat conduction problem with no initial filtering procedure. *Comput. Math. Appl.* **19**, No. 10 (1990), 35–50.

Murio, D. A., and Hinestroza, D. On the space marching solution of the inverse heat conduction problem and the identification of the initial temperature distribution. *Comput. Math. Appl.* **25**, No. 4 (1993), pp. 55–63.

Murio, D. A., and Paloschi, J. R. Combined mollification–future temperatures procedure for solution of inverse heat conduction problem. *J. Comput. Appl. Math.* **23** (1988), 235–244.

Murio, D. A., and Roth, C. C. An integral solution for the inverse heat conduction problem after the method of Weber. *Comput. Math. Appl.* **15** (1988), 39–51.

Muzzy, R. J., Avila, J. H., and Root, R. E. Determination of transient heat transfer coefficients and the resultant surface heat flux from internal temperature measurements. General Electric Report GEAP-20731, 1975.

Nababan, S., and Teo, K. L. Necessary conditions for optimality of Cauchy problems for parabolic, partial delay-differential equations. *J. Optim. Theory Appl.* **34**, No. 1 (1981), 117–155.

Natterer, F. The finite element method for ill-posed problems. *RAIRO Anal. Numér.* **11**, No. 3 (1977), 271–278.

Ohya, Y. Sur l'unicité du prolongement des solutions pour quelques équations differentielles paraboliques. *Proc. Japan Acad.* **37** (1961), 358–362.

Palancz, B., and Szabolcs, G. Solution of inverse heat conduction problem by orthogonal collocation techniques. *Acta Tech.*, *Acad. Sci. Hungar.* **101**, No. 3 (1988), 323–337.

Pankratov, B. M. Methods of analyzing the thermal conditions of engineering systems and inverse heat and mass transfer problems. *J. Engrg. Phys.* **56**, No. 3 (1989), 241–243.

Papukashvili, N. R. Solution of an inverse problem for the heat equation. *Tbiliss. Gos. Univ. Inst. Prikl. Mat. Trudy* **33** (1989), 112–124 (Russian).

Pasquetti, R., and Niliot, C. L. Boundary element approach for inverse heat conduction problems. Applications to a bidimensional transient numerical experiment. *Numer. Heat Transfer* **20**, No. 2 (1991), 169–189.

Payne, L. E. *Improperly Posed Problems in Partial Differential Equations.* SIAM, Philadelphia, 1975.

Payne, L. E. Improved stability estimates for classes of ill-posed Cauchy problems. *Appl. Anal.* **19** (1985), 63–74.

Pham-Lo'i-Vu. Weak Cauchy problems for diffusion equations. *Inverse Problems* **7**, No. 2 (1991), 299–305.

Pollard, H., and Widder, D. V. Inversion of a convolution transform related to heat conduction. *SIAM J. Math. Anal.* **1** (1970), 527–532.

Pucci, C. Teoremi di esistenza e di unicita per il problema di Cauchy nella teoria delle equazioni lineari a derivate parziali. II. *Atti Accad. Naz. Lincei Rend. Cl. Sci. Fis. Mat. Natur.* (8) **13** (1952), 111–116.

Pucci, C. Studio col metodo delle differenze di un problema di Cauchy relativo ad equazioni a derivate parziali del secondo ordine di tipo parabolico. *Ann. Scuola Norm. Sup. Pisa* (3) **7** (1954), 205–215.

Pucci, C. On the improperly posed Cauchy problems for parabolic equations. In *Symp. Numerical Treatment of Partial Differential Equations with Real Characteristics: Proc. Rome Symp.* (January 1959) organized by the Provisional International Computation Centre. Libreria Eredi Virgilio Veschi, Rome, 1959, pp 140–144.

Pucci, C. Alcune limitazioni per le solutioni di equazioni paraboliche. *Ann. Mat. Pura Appl.* (4) **48** (1959), 161–172.

Randall, J. D. Finite difference solution of the inverse heat conduction problem and ablation. Technical Report, Johns Hopkins University, Laurel, MD, 1976. Proc. 25th Heat Transfer and Fluid Mechanics Institute, University of California, Davis, 1976.

Randall, J. D. Embedding multidimensional ablation problems in inverse heat conduction problems using finite differences. *Sixth Internat. Heat Transfer Conf., Toronto, August 7–11, 1978*. Published by National Resource Council of Canada, Toronto, 1978. Available from Hemisphere, Washington, DC, Vol. 3, pp. 129–134.

Raynaud, M. Détermination du flux surfacique traversant une paroi soumise à un incendie au moyen d'une méthode d'inversion. Laboratoire D'Aérothermique Groupe Echanges Thermiques Université Pierre et Marie Curie, Paris, August 1983.

Raynaud, M. Some comments on the sensitivity to sensor location of inverse heat conduction problems using Beck's method. *Internat. J. Heat Mass Transfer* **29**, No. 5 (1986), 815–817.

Raynaud, M. Comparison of space marching finite difference technique and function minimization technique for the estimation of the front location in nonlinear melting problems. *Fifth Symp. Control of Distributed Parameter Systems*, A. El Jai and M. Amoroux (eds.). Université de Perpignan, France, 1989, pp. 215–220.

Raynaud, M., and Beck, J. V. Methodology for comparison of inverse heat conduction methods. *Trans. ASME* **110** (1988), 30–37.

Raynaud, M., and Bransier, J. A new finite difference method for nonlinear inverse heat conduction problem. *Numer. Heat Transfer* **9**, No. 1 (1986), 27–42.

Reemtsen, R., and Kirsch, A. A method for the numerical solution of the one-dimensional inverse Stefan problem. *Numer. Math.* **43** (1984), 253–273.

Reinhardt, H. J. Numerical method for the solution of two-dimensional inverse heat conduction problem. *Internat. J. Numer. Methods in Engrg.* **32**, No. 2 (1991), 363–383.

Reinhardt, H.-J., and Valensia, L. The numerical solution of inverse heat conduction problems with application to reactor technology. In *Structural Mechanics in Reactor Technology (Ninth SMLRL Conf., Lausanne, 1987)*, F. H. Wittmann (ed.). Vol. B, pp. 25–29.

Riganti, R. A solution technique for random and nonlinear inverse heat conduction problems. *Math. Comput. Simulation* **33**, No. 1 (1991), 51–64.

Romanovskii, M. R. Mathematical modeling of experiments with the help of inverse problems. *J. Engrg. Phys.* **57**, No. 3 (1990), 1112–1117.

Sacadura, J. F., and Osman, T. T. Emissivity estimation through the solution of an inverse heat conduction problem. *J. Thermophys. Heat Transfer* **4**, No. 1 (1990), 86–91.

Sachs, E. A parabolic control problem with a boundary condition of the Stefan–Boltzmann type. *Z. Angew. Math. Mech.* **58**, No. 10 (1978), 443–449 (English, German, and Russian summaries).

Schmidt, E. J. P. G. The "bang–bang" principle for the time-optimal problem in boundary control of the heat equation. *SIAM J. Control Optim.* **18**, No. 2 (1980), 101–107.

Schmidt, E. J. P. G. Boundary control for the heat equation with steady-state targets. *SIAM J. Control Optim.* **18**, No. 2 (1980), 145–154.

Scott, E. P., and Beck, J. V. Analysis of order of the sequential regularization solutions of inverse heat conduction problems. *J. Heat Transfer, Trans. ASME* **111**, No. 2 (1989), 218–224.

Seidman, T. I. A well-posed problem for the heat equation. *Bull. Amer. Math. Soc.* **80** (1974), 901–902.

Seidman, T. I. Boundary observation and control for the heat equation. In *Calculus of Variations and Control Theory (Proc. Symp., Math. Res. Center, Univ. Wisconsin, Madison, WI, 1975*; dedicated to Laurence Chosholm Young on the occasion of his 70th birthday), Academic, New York, 1976, pp. 321–351.

Seidman, T., and Eldén, L. An "optimal filtering" method for the sideways heat equation. *Inverse Problems* **6** (1990), 684–696.

Shoji, M., and Ono, N. Application of the boundary element to the inverse problem of heat conduction. *Nippon Kikai Gakkai Ronbunshu, B. Hen* **54**, No. 506 (1988), 2893–2900 (Japanese).

Smulev, I. I. Bounded solutions of boundary-value problems without initial conditions for parabolic equations and inverse boundary-value problems. *Dokl. Akad. Nauk. SSSR* **142** (1962), 46–49 (Russian).

Sparrow, E. M. Haji-Sheikh, A., and Lundgren, T. S. The inverse problem in transient heat conduction. *Trans. ASME Ser. E. J. Appl. Mech.* **31** (1964), 369–375.

Stecher, M. Integral operators and the noncharacteristic Cauchy problem for parabolic equations. *SIAM J. Math. Anal.* **6** (1975), 796–811.

Stolz, G., Jr. Numerical solutions to an inverse problem of heat conduction for simple shapes. *J. Heat Transfer* **82** (1960), 20–26.

Talenti, G. Un problema di Cauchy. *Ann. Scuola Norm. Sup. Pisa* (4) **18** (1964), 165–186.

Talenti, G. Sui problemi mal posti. *Boll. Un. Mat. Ital.* (5) **15-A** (1978), 1–29.

Talenti, G., and Vessella, S. A note on an ill-posed problem for the heat equation. *J. Austral. Math. Soc. Ser. A* **32**, No. 3 (1982), 358–368.

Tandy, D. F., Trujillo, D. M., and Busby, H. R. Solution of inverse heat conduction problems using an eigenvalue reduction technique. *Numer. Heat Transfer* **10**, No. 6 (1986), 597–617.

Tikhonov, A. N., and Arsenin, V. Y. Solutions of ill-posed problems. V. H. Winston and Sons, Washington, DC, 1977.

Tikhonov, A. N., and Glasko, V. V. Methods of determining the surface temperature of a body. *Z. Vychisl. Mat. i Mat. Fiz.* **7** (1967), 910–914.

Tsoi, P. V., Yusunov, S. Y., and Korpeev, N. R. Direct and inverse heat conduction problems for a semi-infinite rod for a partial outflow of heat through the surface. *J. Engrg. Phys.* **47**, No. 1 (1985), 860–864.

Tsutsumi, A. A remark on the uniqueness of the noncharacteristic Cauchy problem for equations of parabolic type. *Proc. Japan Acad.* **41** (1965), 65–70.

Twomey, S. On the numerical solution of Fredholm integral equations of the first kind by the inversion of the linear system produced by quadrature. *J. Assoc. Comput. Mach.* **10** (1963), 97–101.

Twomey, S. The application of numerical filtering to the solution of integral equations encountered in direct sensing measurements. *J. Franklin Inst.* **279** (1965), 95–109.

Vasil'eva, V. N. On the problem of reconstructing the boundary conditions for the heat equation. In *Optimization Methods and Operations Research, Applied Mathe-*

matics (Russian). Akad. Nauk. SSSR Sibersk. Otdel. Sibersk. Energet. Inst., Irkutsk, 1976, pp. 105–114, 189 (Russian).

Vavilov, V. P., and Finkel'shtein, S. V. Two approaches to the solution of the unidimensional inverse problem of thermal quality control. *Soviet J. Nondestructive Testing* **25**, No. 4 (1989), 284–287.

Walter, G. G. An alternative approach to ill-posed problems. *J. Integral Equations Appl.* **1** (1989), 287–301.

Weber, C. F. Analysis and solution of the ill-posed inverse heat conduction problem. *Internat. J. Heat Mass Transfer* **24** (1981), 1783–1792.

Weiland, E., and Babary, J. P. Comparative study for a new solution to the inverse heat conduction problem. *Comm. Appl. Numer. Methods* **4**, No. 5 (1988), 687–689.

Williams, S. D., and Curry, D. M. An analytical experimental study for surface heat flux determination. *J. Spacecraft* **14** (1977), 632–637.

Winther, R. Error estimates for a Galerkin approximation of a parabolic control problem. *Ann. Mat. Pura Appl.* (4) **117** (1978), 173–206.

Woo, K. C., and Chow, L. C. Inverse heat conduction by direct inverse Laplace transform. *Numer. Heat Transfer* **4** (1981), 499–504.

Woodbury, K. Effect of thermocouple sensor dynamics on surface heat flux predictions obtained via inverse heat transfer analysis. *Internat. J. Heat Mass Transfer* **33**, No. 12 (1990), 2641–2649.

Yamabe, H. A unique continuation theorem of a diffusion equation. *Ann. Mat.* (2) **69** (1959), 462–466.

Yoshimura, T., and Ikuta, K. Inverse heat conduction problem by finite element formulation. *Internat. J. Systems Sci.* **16** (1985), 1365–1376.

Zabaras, N. Inverse finite element techniques for the analysis of solidification processes. *Internat. J. Numer. Methods in Engrg.* **29**, No. 7 (1990), 1569–1587.

Zabaras, N., and Liu, J. C. An analysis of two-dimensional linear inverse heat transfer problems using an integral method. *Numer. Heat Transfer* **13** (1990), 527–533.

Zabaras, N., and Ruan, Y. Deforming finite element method analysis of inverse Stefan problems. *Internat. J. Numer. Methods in Engrg.* **28**, No. 2 (1989), 295–313.

Zabaras, N., and Ruan, Y. Moving and deforming finite-element simulation of two-dimensional Stefan problems. *Comm. Appl. Numer. Methods* **6**, No. 7 (1990), 495–506.

Zabaras, N., Mukherjee, S., and Richmond, D. An analysis of inverse heat transfer problems with phase changes using an integral method. *ASME J. Heat Transfer* **110** (1988), 554–561.

SUBJECT INDEX